Coastal and Estuarine Studies

Managing Editors:
Malcolm J. Bowman   Richard T. Barber
Christopher N.K. Mooers   John A. Raven

Coastal and Estuarine Studies

formerly Lecture Notes on Coastal and Estuarine Studies

33

M.L. Khandekar

Operational Analysis and Prediction of Ocean Wind Waves

Springer-Verlag

New York Berlin Heidelberg London Paris Tokyo Hong Kong

Author

M.L. Khandekar
Atmospheric Environment Service
4905 Dufferin Street
Downsview, Ontario M3H 534, Canada

ISBN-13: 978-1-4613-8954-5   e-ISBN-13: 978-1-4613-8952-1
DOI: 10.1007/978-1-4613-8952-1

PREFACE

This monograph is an attempt to compile the present state of knowledge on ocean wave analysis and prediction. The emphasis of the monograph is on the development of ocean wave analysis and prediction procedures and their utility for real-time operations and applications. Most of the material in the monograph is derived from journal articles, research reports and recent conference proceedings; some of the basic material is extracted from standard text books on physical oceanography and wind waves.

Ocean wave analysis and prediction is becoming an important activity in the meteorological and oceanographic services of many countries. The present status of ocean wave prediction may be comparable to the status of numerical weather prediction of the mid-sixties and early seventies when a number of weather prediction models were developed for research purposes, many of which were later put into operational use by meteorological services of several countries. The increased emphasis on sea-state analysis and prediction has created a need for a ready reference material on various ocean wave analysis and modelling techniques and their utility. The present monograph is aimed at fulfilling this need. The monograph should prove useful to the ocean wave modelling community as well as to marine forecasters, coastal engineers and offshore technologists. The monograph could also be used for a senior undergraduate (or a first year graduate) level course in ocean wave modelling and marine meteorology.

The operational running of an ocean wave model requires an appropriate wind field which is generally extracted from an operational numerical weather prediction model; this, in my opinion, has helped develop a better communication between operational meteorologists and applied oceanographers. Future development on wave model initialization, as discussed briefly in the last Chapter of the monograph, is likely to bring together researchers from various disciplines including meteorology, oceanography, satellite remote sensing and microwave technology. This may provide closer co-operation among various disciplines which could help produce improved solutions to problems of mutual interest.

Atmospheric Environment Service
Downsview, Ontario, Canada
June 1989

Madhav L. Khandekar

ACKNOWLEDGEMENTS

I wish to express my sincere thanks to Dr. Malcolm Bowman, Managing Editor, for his encouragement and continuing support throughout the preparation of this monograph. Several discussions with a number of my colleagues and with other research workers have provided a valuable input into the monograph. In particular, I wish to acknowledge fruitful discussions with Vince Cardone (Oceanweather Inc., Cos Cob, U.S.A.), Mark Donelan (National Water Research Institute, Burlington, Canada), Steve Peteherych (Atmospheric Environment Service, Downsview, Canada), Don Resio (Offshore and Coastal Technologies, Inc., Vicksburg, U.S.A.), Rachel Stratton (British Meteorological Office, Bracknell, U.K.) and Liana Zambresky (European Centre for Medium-Range Weather Forecasts, Reading, U.K.).

My affiliation with the Atmospheric Environment Service (AES) has provided me with an opportunity to get involved with ocean wave modelling; this has provided an impetus towards preparation of this monograph. I wish to thank Mr. Alan Bealby, Chief, Forecast Research Division, AES for making available necessary facilities for the preparation of the monograph. The drafting division of the AES provided valuable help in drafting many of the diagrams of the monograph. The assistance of Dr. Ron Wilson, Director, Marine Environmental Data Service, Ottawa and his co-workers in providing wave data and wave plots is gratefully acknowledged.

Mrs. Pearl Burke of the Canadian Climate Centre, AES deserves special thanks for expertly typing several draft versions of the monograph including the final version in the camera-ready format. Thanks are also due to Miss Sandra Scott, a summer student from the University of Waterloo for her help in proof-reading the manuscript and checking the mathematical symbols. Finally, I wish to express my gratitude to my wife Shalan for her intellectual and emotional support during the preparation of the monograph.

TABLE OF CONTENTS

TABLE OF CONTENTS (Cont'd)

CHAPTER 1
INTRODUCTION

Since the pioneering development of wave forecasting relations by Sverdrup and Munk(1947), significant advances have been made in ocean wave prediction. At present, spectral wave models based on the energy balance equation of a wave field are used in operational modes in many areas of world oceans. Several theoretical and observational studies over the last forty years have provided a reasonable quantitative formulation of the various source terms in the energy balance equation, the cornerstone of modern spectral ocean wave models. Despite these advances, a remark made by Dr. F. Ursell in his well-known paper 'Wave Generation by Wind' (Ursell, 1956) still applies in the following modified form: "wind blowing over water surface generates waves in the water by physical processes which cannot be regarded as completely known at present".

There is a hierarchy of waves at any given point in an ocean. In general, five basic types of waves namely sound, capillary, gravity, inertial and planetary waves have been identified which can and do occur in an ocean with five basic restoring forces all acting simultaneously to produce more complicated mixed types of waves and oscillations. The relative importance of each restoring force in any particular situation depends upon the properties of the medium, the geometry of the ocean basin in question and the frequency and wave-length of oscillation. The energy spectrum of oceanic variability can be schematically shown in Figure 1.1 where the energy variation along the ordinate is shown in arbitrary units. Starting from the short capillary waves (wavelengths of the order of only a couple of cm or so), one encounters a narrow band of wind-induced surface gravity waves with a period ranging from 1 to 10 seconds; longer-period (25 s or more) surface gravity waves are sometimes generated in response to sustained meteorological forcing (ex. a sequence of weather disturbances in the central Pacific producing long-period surface waves further downstream in the northeast Pacific, offshore northwestern United States); seismic activities at the ocean floor can create very long-period forced gravity waves commonly known as tsunamis. The tides are another type of forced gravity waves. At very long periods, gravity loses its dominant role to differential rotation effects of the earth and the surface waves become planetary waves and manifest themselves as slowly drifting large-scale current systems. The present

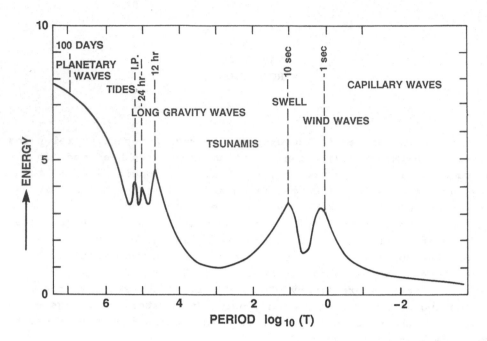

Figure: 1.1: Schematic energy spectrum of ocean variability, showing
different types of waves occurring in the ocean. Energy variation
along the ordinate is shown in arbitrary units. I.P. denotes inertial
period which is ∿ 35 hr. (corresponding to a latitude of 20°) in this
Figure. (from LeBlond and Mysak, 1978).

monograph is concerned with the wind-induced surface gravity waves
which are most commonly seen along a sea shore and which have the
maximum impact on human activity.

One of the most fascinating features of the surface of a sea
is the innumerable waves of different shapes and sizes at any given
time; the sea surface is rarely ever calm. Even under calm wind conditions, the sea surface might undulate with smooth swells whose
source is a storm which may have occurred a few days before and hund-
reds of kilometers away. With a light breeze, an intricate pattern of
capillary waves is formed; these capillary waves (or wavelets) are
only a few centimeters in length and a few millimetres in height and
they form a remarkably regular diamond shaped pattern as they wrinkle
the sea surface. As the wind continues to blow, the capillary waves
grow in size increasing in their wavelengths and become surface
gravity waves where gravity is the most dominating restoring force.
Field observations suggest that for wind speeds of up to 1 m s^{-1} (approximately 2 knots), capillary waves predominate the ocean surface
while a somewhat stronger breeze transforms these capillary waves into
gravity waves. These gravity waves continue to grow as long as the

energy income of the waves from all sources is greater than the energy dissipation by whatever possible means, wave breaking and turbulent water motion being the principal ones. At first, waves grow both in length and height; if the waves reach their height limit and there is more energy available, the growth is then predominantly in length.

How high the waves will grow will depend upon the wind force or the wind speed; it also depends upon the wind duration, i.e. how long the wind blows and on the over-water fetch, i.e. the distance travelled by the wind over water. In other words, wind needs not only strength, but time as well as elbowroom to produce higher and longer waves. As an example, the Hibernia area (approx. 46°N, 48°W) off Newfoundland in the Canadian Atlantic often experiences severe winter storms (during January to March) which generally move along a southwest-northeast track. The effective fetch in such a storm rarely exceeds 750 km (about 400 nautical miles). Over this fetch, a wind speed of 25 m s^{-1} ($\sim$ 50 knots) often produces wave heights in excess of 10 m at the Hibernia area. Extreme storm-sea conditions (wave heights in excess of 15 m) often develop over the Cape Horn area at the southern tip of South America where strong winds can blow over long fetches of unbroken oceans of the Antarctic.

In general, longer waves with faster speeds also have longer periods. The wave period (T), wavelength (L) and the wave phase speed (c) for deep-water waves are related by an equation,

$$c = \sqrt{gL/2\pi} \qquad (1.1)$$

here g is the gravitational acceleration. Using the expression (1.1), the following Table giving values of T, L and c can be prepared.

Table 1.I Values of period, length and speed for deep-water waves

Period, T(s)	2	4	8	16
Length, L(m)	6.2	25.0	99.8	399.3
Phase speed, c(m s^{-1})	3.1	6.2	12.5	25.0

It can be seen from this table that the speed of the deep-water waves depend on their wavelengths and also on their periods, i.e. they are dispersive waves. Thus if a number of waves of different wavelengths are generated simultaneously over a given area in the ocean, the longer waves will move ahead of the shorter ones and will be observed first at a distant point from the source area. The long waves are thus the 'forerunners' of an incoming storm, a fact recognized in folklore and by some of the primitive people of tropical islands.

The fastest ocean surface waves are not wind waves but tsunamis or seismic sea waves. A tsunami generated after the well-known volcanic eruption of Krakatoa in Indonesia (in 1883) was estimated to have travelled at a speed of about 150 m s^{-1}. The tsunami which reached the Hawaiian islands on 1 April 1946 came from the Aleutian chain of islands, travelling at an estimated speed of about 215 m s^{-1}.

1.1 Scope of the Monograph

In Chapter 2, brief details of the basic wave dynamics are presented while Chapter 3 provides a summary of important studies on wave generation, propagation and dissipation. These two Chapters are intended to provide a ready reference material and a suitable basis for various wave prediction models presented in later Chapters. It must be emphasized here that only a skeleton account of basic wave dynamics is presented in Chapter 2; for a detailed treatment on wave dynamics, the reader is referred to the classical book on wind waves by Kinsman(1965) or a recent and a more advanced book on waves in the ocean by LeBlond and Mysak(1978).

In Chapter 4, the pioneering wave prediction technique of Sverdrup and Munk is summarized; this is followed by some details of the wave spectrum method developed by Pierson, Neumann and James (1955). Chapter 5 provides a physical basis for the modern spectral wave models and summarizes a number of spectral wave models developed in the last twenty-five years. Chapter 6 provides a framework for wave modelling in shallow water and discusses the inclusion of shallow-water effects in operational wave prediction models.

Chapter 7 is devoted to wave model validation and includes a number of studies reported in the last fifteen years on evaluation of various models; the Chapter also includes results of a wave model intercomparison study initiated recently by the Atmospheric Environment Service (AES), Canada. These studies have provided a certain degree of confidence in the use of various products generated by the wave models.

Chapter 8 discusses an important aspect of wave analysis and prediction namely wind specification for operational wave models. The boundary layer model of Cardone(1969) is briefly reviewed which is followed by a discussion on wind specification procedures developed for operational running of wave models. Chapter 9 discusses the

operational wave analysis and interpretation of wave records; the
Chapter also discusses the wave climatology and related database that
has evolved from the wave model hindcasting projects initiated in many
countries. The Chapter concludes with a section giving brief details
on real-time wave analysis and prediction in Canada, U.S.A., Europe
and elsewhere.

The last Chapter summarizes the present status of wave prediction and discusses future wave modelling efforts with emphasis on the
use of satellite-sensed wind and wave data.

REFERENCES

Cardone, V.J., 1969: Specification of the wind distribution in the
marine boundary layer for wave forecasting. Geoph. Science Laboratory,
TR-69-1, School of Eng. and Sciences, New York University, 118 pp.,
Dec. 1969.

Kinsman, B., 1965: Wind waves: their generation and propagation on the
ocean surface. Prentice Hall, U.S.A., 676 pp. (Dover Publications,
U.S.A., 1984).

LeBlond, P. and L.A. Mysak, 1978: Waves in the ocean. Elsevier, The
Netherlands, 602 pp.

Pierson, W.J., G. Neumann and R.W. James, 1955: Practical methods for
observing and forecasting ocean waves by means of wave spectra and
statistics. H.O. Pub. 603, U.S. Navy Hydrographic Office, Washington,
D.C., 284 pp.

Sverdrup, H.U. and W.H. Munk, 1947: Wind sea and swell: Theory of
relations for forecasting. H.O. Pub. 601, U.S. Navy Hydrographic
Office, Washington, D.C., 44 pp.

Ursell, F., 1956: Wave generation by wind. Surveys in mechanics,
Cambridge University Press, U.K., 216-249.

For the sake of simplicity, we consider a sinusoidal form of
Figure 2.1 to represent a typical wave generated on the free surface
of a sea. It must be emphasized here that an ocean surface rarely ever
exhibits a simple and a singular waveform as displayed in Fig. 2.1.
With this waveform are associated various wave parameters which can be
defined in the following standard notations:

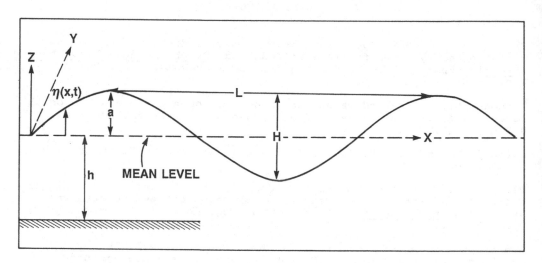

Figure 2.1: A wave profile with various parameters in standard
notations.

a: wave amplitude
H: wave height from crest to trough (H=2a)
L: wave length
$\eta(x,t)$: instantaneous vertical displacement of water surface above
    mean water level.
k: wave number = $2\pi/L$
T: wave period (the time interval between the occurrence of successive
    troughs or crests at a given fixed position)
c: phase speed = L/T
f: wave frequency = $1/T$; σ : wave angular frequency = $2\pi/T$; $\sigma=kc$
h: water depth measured from the mean water level.
δ: wave steepness = H/L. According to Stokes' theory, maximum value of
    H/L is 1/7.
β: wave age = c/U; here U is the horizontal wind speed. As the wave
    grows with the wind speed, it increases in wave length which leads
    to an increase in its phase speed c. Thus β, the wave age grows as
    the wave grows.

A point on the wave profile is given by a variable η which can be expressed using a generalized sinusoidal form

$$\eta(x,t) = a \cos(kx - \sigma t + \epsilon) \tag{2.1}$$

Here ε is the phase shift, t is time and x is the distance measured in the direction of propagation of the wave. Equation(2.1) represents a wave propagating along the x-axis and whose properties do not change along a line perpendicular to the x-axis, i.e. along the y axis, for example. A more generalized expression than (2.1) involving x and y can be used if a wave is assumed to propagate along any arbitrary line. (It may be noted that if the minus sign is changed to a plus, the wave would propagate the other way). Further, using the relationship σ = kc and assuming ε = o, we can re-write (2.1) as

$\eta(x,t) = a \cos\{k(x - ct)\}$. This equation represents a wave propagating along the x-axis with a phase speed c. In the following section, we will obtain the phase speed solution of an ocean surface wave described by the above equation, using elementary principles.

2.1 Phase Speed Solution of a Surface Wave

The most convenient starting point for obtaining the phase speed solution is to express the governing equations of the motion in two horizontal (u,v) and one vertical (w) components of the total fluid velocity $\bar{U}$. For simplicity, we assume incompressible, inviscid fluid flow on a non-rotating earth. We further assume that the wave properties do not change along the y-axis, so that variations with respect to y are neglected. With these assumptions, the governing equations can be written in standard notations as follows:

$$\frac{\partial u}{\partial t} + u\frac{\partial u}{\partial x} + w\frac{\partial u}{\partial z} = -\frac{1}{\rho}\frac{\partial p}{\partial x}$$

$$\frac{\partial w}{\partial t} + u\frac{\partial w}{\partial x} + w\frac{\partial w}{\partial z} = -\frac{1}{\rho}\frac{\partial p}{\partial z} -g \tag{2.2}$$

$$\frac{\partial u}{\partial x} + \frac{\partial w}{\partial z} = 0$$

In (2.2), the first two equations are the x- and the z-components of the equation of motion and the third equation is the continuity equation for an incompressible flow. In these equations, p is the pressure, ρ is the fluid density, g is the acceleration due to gravity and the other symbols are already defined.

There exists an important class of fluid motion called irrotational fluid flow which can be defined by a mathematical condition

∇ x $\bar{U}$ (or Curl $\bar{U}$) = 0; here $\bar{U}$ is the velocity vector. For an irrotational flow, we can define a velocity potential ϕ which allows us to express u and w as

$$u = -\frac{\partial \phi}{\partial x} \; ; \quad w = -\frac{\partial \phi}{\partial z} \tag{2.3}$$

Assume wave motion to be irrotational and substitute (2.3) in (2.2); integrate the first two equations and simplify to obtain

$$gz - \frac{\partial \phi}{\partial t} + \frac{1}{2}\left\{\left(\frac{\partial \phi}{\partial x}\right)^2 + \left(\frac{\partial \phi}{\partial z}\right)^2\right\} + \frac{p}{\rho} = \text{const.}$$

$$\frac{\partial^2 \phi}{\partial x^2} + \frac{\partial^2 \phi}{\partial z^2} = 0. \tag{2.4}$$

The first equation of (2.4) is one of the many forms of Bernoulli's equation, while the second is the well-known Laplace's equation often expressed as $\nabla^2 \phi = 0$. In order to solve (2.4), appropriate boundary conditions must be prescribed. For the problem under consideration, there are two boundaries to be considered, one the air-water interface (or the free surface) and the other, the rigid bottom boundary. At the air-water interface, the curve that forms the boundary can be defined by z = η(x,t); along this boundary, the fluid is assumed to move tangentially so that it always remains at the boundary. This condition is described mathematically by the requirement that $\frac{d}{dt}[z - \eta(x,t)] = 0$; this can be simplified to write as

$$\frac{\partial \phi}{\partial z} + \frac{\partial \eta}{\partial t} = \frac{\partial \phi}{\partial x}\frac{\partial \eta}{\partial x} \qquad \text{for } z = \eta(x,t) \tag{2.5}$$

Equation (2.5) is identified as the kinematic boundary condition. Also, on this boundary, the pressure must be constant as no pressures due to the air on the water are assumed to be acting. This reduces the Bernoulli equation to

$$gz - \frac{\partial \phi}{\partial t} + \frac{1}{2}\left[\left(\frac{\partial \phi}{\partial z}\right)^2 + \left(\frac{\partial \phi}{\partial z}\right)^2\right] = \text{const.} \qquad \text{for } z = \eta(x,t) \tag{2.6}$$

At the bottom boundary where the water depth h is assumed constant, the impermeable bottom requires that the normal component of the fluid motion is zero; this gives

$$w = -\frac{\partial \phi}{\partial z} = 0 \qquad \text{at } z = -h \tag{2.7}$$

Finally, the Laplace's equation must be satisfied throughout the fluid domain, so we have

$$\frac{\partial^2 \phi}{\partial x^2} + \frac{\partial^2 \phi}{\partial z^2} = 0 \qquad \text{for } -h \le z \le \eta(x,t) \tag{2.8}$$

Equations (2.5) to (2.8) are to be solved to obtain the solution for η and ϕ. A closer inspection reveals that these equations contain non-linear terms and further the derivatives of ϕ are to be evaluated at the surface $z = \eta(x,t)$ whose solution is not yet known. This makes the solution of these equations an extremely difficult problem. In order to resolve this mathematical difficulty, the following assumptions and simplifications are made:

$\underline{1}$. The potential function ϕ and its derivatives are expressed in a Taylor series expansion about $z = 0$, so that the boundary conditions at the free surface can be evaluated at $z = o$ instead of at $z = \eta(x,t)$.

$\underline{2}$. A steady-state fluid flow is assumed by introducing a factor cx to the velocity potential where c is the phase speed whose solution is sought. With a steady-state assumption, all the derivatives with respect to time are eliminated.

$\underline{3}$. The amplitude of the surface disturbance is assumed small compared to the wavelength (L) or the depth (h), i.e. $ak \ll 1$ and $\frac{H}{L} \ll 1$.

With these assumptions, the unknown functions (ϕ, η and c) in the equations (2.5) through (2.8) are expanded in powers of a small dimensionless parameter ak and terms containing the first power of ak (first- order terms) are collected to obtain the following linearized set of equations (see Neumann and Pierson, 1966):

$$\frac{\partial \phi}{\partial z} - c\frac{\partial \eta}{\partial x} = 0 \qquad \text{at } z = o$$

$$g\eta + c\frac{\partial \phi}{\partial x} = 0 \qquad \text{at } z = o$$

$$\frac{\partial^2 \phi}{\partial x^2} + \frac{\partial^2 \phi}{\partial z^2} = 0 \qquad \text{for } z < \eta \tag{2.9}$$

$$\frac{\partial \phi}{\partial z} = 0 \qquad \text{at } z = -h$$

In (2.9), the variables ϕ, η and c denote the first-order approximations (for simplicity, the subscript 1 denoting the first-order approximation has been dropped from these variables). For this linearized set of equations we seek a wave form solution expressed as $\eta = a$ cos kx. Substitute this solution in (2.9) and use boundary conditions to obtain an expression for the potential function as

$$\phi = \frac{- ca \cosh k(z + h)}{\sinh kh} \sin kx$$

Using this expression of ϕ in the second equation of (2.9) yields a condition

$$c^2 = \frac{g}{k} \tanh kh \qquad (2.10)$$

Equation (2.10) is the phase speed solution of a surface wave obtained using the basic fluid dynamical equations and principles. In deriving (2.10), we used three simplifying assumptions to reformulate the set of equations as discussed above. These assumptions and the related procedure form the backbone of a powerful mathematical procedure called the perturbation technique which assume the total fluid flow consisting of two quantities, an equilibrium value and a perturbation value which is at least one order of magnitude smaller than the equilibrium value. This assumption allows the governing equations of motion to be transformed into a linearized form. In the preceeding derivation, we have not explicitly specified the magnitude of the equilibrium flow but have tacitly assumed that the equilibrium flow is at rest, which means that the horizontal velocity components (u,v) are zero everywhere in the equilibrium state. It is possible to consider a nonzero equilibrium state so that we can express the independent variables u, w, p and ρ as:

$$(u, w, p, \rho) = (U_o, o, p_o, \rho_o) + (u', w', p', \rho') \qquad (2.11)$$
$$\text{Total} \qquad \text{Equilibrium} \quad + \quad \text{Perturbation}$$

Here the equilibrium (or the undisturbed) state of the fluid can be described analytically by the following conditions:

$$U_o = \text{constant} ; \quad \frac{\partial p_o}{\partial z} = -g\rho_o \qquad (2.12)$$

Equation (2.12) states that the fluid in equilibrium state is moving with a constant horizontal speed U_o and is in hydrostatic balance. We can now rewrite the governing equations (2.2) using the equilibrium and the perturbation quantities and applying the perturbation conditions (i.e. neglecting terms containing product of perturbation quantities) we obtain

$$\frac{\partial u'}{\partial t} + U_o \frac{\partial u'}{\partial x} = -\frac{1}{\rho_o} \frac{\partial p'}{\partial x}$$

$$\frac{\partial w'}{\partial t} + U_o \frac{\partial w'}{\partial x} = -\frac{1}{\rho_o} \frac{\partial p'}{\partial z} - g \qquad (2.13)$$

$$\frac{\partial u'}{\partial x} + \frac{\partial w'}{\partial z} = o$$

The set of equations (2.13) is a linearized set expressed in terms of a basic flow U_o and perturbation quantities u', w' and p'. It may be noted here that the perturbation state of the fluid flow is considered nonhydrostatic, unlike the equilibrium state which is assumed to be in hydrostatic balance as defined by eq.(2.12). In order to solve the set of eq.(2.13), the boundary conditions discussed earlier are expressed in the following form:

$$w' = o \quad \text{at} \quad z = 0 ; \qquad \text{bottom boundary condition}$$

$$\frac{d}{dt}(p) = \frac{d}{dt}(p_0 + p') = 0 \qquad \text{at the free surface} \tag{2.14}$$

Assume wave form solutions for the perturbation quantities u', w' and p'; for ex. $u' = \psi(z) \exp[ik(x - ct)]$, where $\psi(z)$ is an arbitrary function, k is the wave number and c is the phase speed. Substitute these wave form solutions in the linearized set (2.13) and apply the boundary conditions. The second boundary condition is rewritten as

$$\frac{d}{dt}(p_0 + p') = o \rightarrow \frac{\partial p'}{\partial t} + U_0 \frac{\partial p'}{\partial x} + w' \frac{\partial p_0}{\partial z} = o \quad \text{and is applied at}$$

$z = h + h' \approx h$. A little manipulation yields

$$(U_0 - c)^2 = \frac{g}{k} \tanh kh$$

or
$$c = U_0 \pm \sqrt{\frac{g}{k} \tanh kh} \tag{2.15}$$

Equation (2.15) gives the phase speed of a surface gravity wave in a fluid medium of constant depth h and moving with a constant speed U_0. This equation can be compared with eq.(2.10) which obtains the same formula under the assumption that the fluid medium is at rest initially. The preceeding development provides a different approach to the problem of solving a set of nonlinear equations governing a fluid flow and eq.(2.15) may be interpreted as a general expression for the phase speed of a surface gravity wave. This approach is commonly used in atmospheric dynamics and numerical weather prediction where surface gravity waves can often develop in a zonal atmospheric flow (see for ex. Haltiner and Williams, 1980).

Equation (2.10) or (2.15) is obtained under the assumption that the fluid depth h is constant. Two special cases can be derived from these equations as discussed below:

Case 1: Deep-water waves

If the fluid depth $h > 0.5L$; in this case $\tanh \frac{2\pi h}{L} \approx 1$ and using the generalized expression (2.15) we have,

$$c = U_0 \pm \sqrt{\frac{gL}{2\pi}} \tag{2.16}$$

Equation (2.16) is known as the Stokes' formula for deep-water waves; this formula was derived by Stokes in 1847. Assuming a zero basic speed of the fluid medium ($U_o = o$) and using $c = \frac{L}{T}$, the Stokes' formula can be rewritten as $L = \frac{g}{2\pi} T^2$; further, using appropriate values for g and π, we have the following deep-water wave relations:

$$L(\text{feet}) \approx 5.12 \ T^2(s)$$
$$L(\text{metres}) \approx 1.56 \ T^2(s) \tag{2.17}$$

Case 2: Long waves in shallow water

If the water depth $h < \frac{L}{200}$, we can write $\tanh \frac{2\pi h}{L} \backsim \frac{2\pi h}{L}$ with the same (percentage) accuracy as we write $\tanh \frac{2\pi h}{L} \backsim 1$ when $h > 0.5L$; this transforms equation (2.15) into

$$c = U_o \pm \sqrt{\frac{gL}{2\pi} \cdot \frac{2\pi h}{L}} = U_o \pm \sqrt{gh} \tag{2.18}$$

This is the well-known formula for shallow-water gravity waves. It may be noted that the solution (2.18) can also be obtained directly from the set of equations (2.13) by considering the perturbation flow to be in hydrostatic equilibrium so that the second equation of (2.13) can be written is $-\frac{1}{\rho_o} \frac{\partial p'}{\partial z} - g = o$; with this modification, the set of eq.(2.13) leads to the shallow-water gravity wave formula given by (2.18).

According to Stokes' formula, waves in deep water travel independently of the water depth and the wave speed depends only upon the wave length. As can be seen from Table 1.1, a 100-m wave travels twice as fast as a 25-m wave thus making deep-water waves dispersive; accordingly, the phase speed formula (2.10) is often referred in literature as the dispersion relationship. In case of shallow water, the wave speed depends only on the depth; thus long waves approaching a uniformly sloping beach at an angle will be refracted so as to arrive almost parallel to the beach. This explains the fact that wavebreakers at the seashore are almost always built parallel to the beach.

The terms deep- and shallow-water waves are relative. Mathematically speaking, it is the ratio $\frac{h}{L}$ which determines how the waves should be classified. For $\frac{h}{L} \geq \frac{1}{2}$, the trigonometric function $\tanh \frac{2\pi h}{L} \geq \tanh \pi = 0.9963$. Using the formula,

$$c = \sqrt{\frac{gL}{2\pi} \tanh \frac{2\pi h}{L}} \quad \text{we can write}$$

$c_d \geq c \geq 0.9963\ c_d$. Here c_d refers to the deep-water waves for which we take $\tanh \frac{2\pi h}{L} \sim 1$. Thus the error in phase speed by using the approximate deep-water formula ($c \sim \sqrt{\frac{gL}{2\pi}}$) whenever the water depth is greater than one-half the wave length is, at most 0.37 percent. From a practical point of view, a five percent error in wave measurement is considered quite acceptable; hence for practical oceanographic work,

$$c \sim \sqrt{\frac{gL}{2\pi}} \qquad \text{when } \frac{h}{L} \geq \frac{1}{4} \qquad \text{deep-water waves}$$

$$(2.19)$$

$$c \sim \sqrt{gh} \qquad \text{when } \frac{h}{L} \leq \frac{1}{20} \qquad \text{shallow-water waves}$$

In (2.19), the basic current U_0 is assumed zero.

Figure 2.2 shows the variation of wave speed versus water depth for different wave lengths. The shallow-water wave speed varies along the straight line through the origin, while the deep-water wave

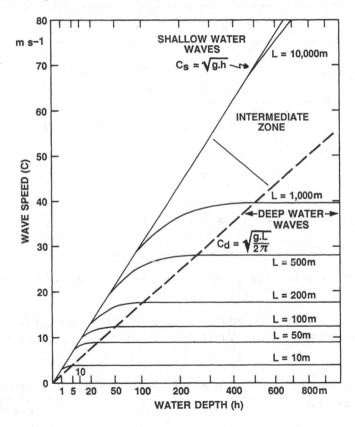

Figure 2.2. Variation of wave speed with water depth for different wave lengths (from Pond and Pickard, 1983)

speed increases for increasing wave lengths as shown by lines almost
parallel to the x-axis.

2.2   Internal and Capillary Waves

The phase speed solution obtained earlier assumes that the
density of the overlying medium (air in case of ocean surface waves)
is negligible in comparison to the density of water and that the pressure on the two sides of the air-water interface is the same. The
first assumption yields a wave solution applicable to a free surface
wave; the second assumption applies to a situation in which waves
developed on the air-water interface have long enough wave lengths
that the surface tension force can be neglected. Both these assumptions can be removed to obtain expressions which can describe internal
and capillary waves.

Figure 2.3 shows a sinusoidal wave at an interface between two
fluids having densities ρ_1 and ρ_2 ($\rho_1 > \rho_2$) so that the fluids are

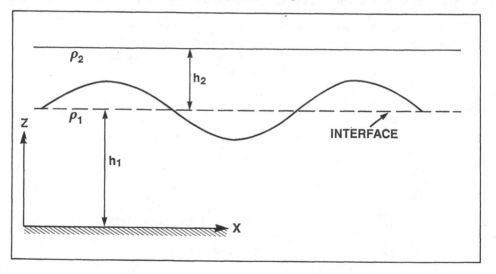

Figure 2.3:   A wave profile at an interface between two fluids having
densities ρ_1 and ρ_2 ($\rho_1 > \rho_2$).

gravitationally stable. The equilibrium surface between the two
fluids is located at $z = h_1$. The following boundary conditions are
prescribed:

1. $w_1 = 0$ at $z = 0$ and $w_2 = 0$ at $z = h_1 + h_2$ kinematic
     $w_1 = w_2$ at $z = h_1$ boundary
                                                                            conditions

2. $\frac{d}{dt}(p_{01} + p_1') = \frac{d}{dt}(p_{02} + p_2')$ dynamic
                                                                            boundary
                                                                            condition

In the above, w_1 and w_2 are vertical velocities in lower and upper layers, p_{01} and p_{02} are equilibrium pressures in the two layers, p_1' and p_2' are the perturbation pressures and h_1 and h_2 are the depths of the two fluids which are considered as bounded by two rigid parallel plates, one at the bottom (where $z = 0$) and the other at the free surface (where $z = h_1 + h_2$). To obtain the phase speed solution, we assume solutions for w_1 and p_1 (and also for w_2 and p_2) which are combinations of exponential and hyperbolic functions. Apply the boundary conditions and simplify to obtain an expression,

$$c^2 = \frac{g}{k} \frac{(\rho_1 - \rho_2)}{\rho_1 \coth kh_1 + \rho_2 \coth kh_2} \qquad (2.20)$$

Equation (2.20) is the phase speed formula for internal waves developed on an interface between two fluids of different densities when surface tension across the interface is neglected. With the inclusion of the surface tension T, the dynamic boundary condition will be modified and the phase speed solution can be rederived as

$$c^2 = \frac{\frac{g}{k}(\rho_1 - \rho_2) + kT}{\rho_1 \coth kh_1 + \rho_2 \coth kh_2} \qquad (2.21)$$

In deriving (2.20) or (2.21), we have tacitly assumed that the two fluids are at rest. If the two fluids move with a certain basic flow, say U, which is either same in both the layers or has different values in upper and lower layers, the phase speed solution can be obtained by modifying the boundary conditions; these solutions are discussed later.

When the fluid surface is produced not by two chemically different fluids but by a sharp density gradient within a single fluid such as may occur at a halocline or a thermocline in the ocean, the surface tension will be negligible and the phase speed will be given by (2.20). Once again, we can consider the deep- and shallow-water approximations for the two fluids. Of the four possible combinations, we consider the following two in some detail since these two combinations are more likely to occur in an oceanic environment.

a. Lower layer deep, upper layer shallow

Use suitable approximations for the hyperbolic functions so that $\coth kh_1 \backsim 1$ when $\frac{h_1}{L} \gg 1$, $\coth kh_2 \backsim \frac{1}{kh_2}$ when $\frac{h_2}{L} \ll 1$. With these approximations, Eq.(2.20) is reduced to,

$$c^2 = \frac{gh_2(\rho_1 - \rho_2)}{\rho_1 kh_2 + \rho_2}$$

If the density difference is small as is generally the case across a thermocline for example, where the difference is about 0.002 gm cm^{-3}(=2 kg m^{-3}), we can write $\rho_1 \sim \rho_2$ in the denominator; further, dividing the numerator and the denominator by ρ_1 we get,

$$c^2 = \frac{gh_2(1 - \rho_2/\rho_1)}{kh_2 + \frac{\rho_2}{\rho_1}} \approx \frac{gh_2(1 - \frac{\rho_2}{\rho_1})}{kh_2 + 1}$$

Further, since $kh_2 \ll 1$, the upper layer being shallow, we have

$$c^2 \approx gh_2 \left(1 - \frac{\rho_2}{\rho_1}\right) = g'h_2 \qquad (2.22)$$

Equation (2.22) is the formula for the phase speed across a thermocline; these waves are non-dispersive and analogous to long surface waves in waters of depth h_2 but with a considerably 'reduced gravity' $g' = g(1 - \rho_2/\rho_1)$. Typically, internal waves have much smaller phase speeds but large amplitudes because of small density differences between the layers. An example of internal waves having periods up to 12 hours and amplitudes from 10 to 100 m or more near the continental shelf-break off the Great Barrier Reef in the Australian northeast coast offshore has been discussed by Pond and Pickard (1983). Large internal waves extending several tens of kilometers have been observed by earth orbiting satellites in the Andaman Sea (southeast of Burma/Thailand border); these giant waves have been mathematically analyzed as solitary waves by Osborne and Burch(1980).

b. **Both upper and lower layer shallow**

Using once again appropriate approximations for the hyperbolic functions, we have

$$c^2 \approx \frac{gh_1h_2(\rho_1 - \rho_2)}{\rho_1h_2 + \rho_2h_1}.$$ Simplifying further when $\frac{\rho_1}{\rho_2} \sim 1$,

we have $c^2 \approx \frac{gh_1h_2}{(h_1 + h_2)}\left(1 - \frac{\rho_2}{\rho_1}\right)$ $\qquad (2.23)$

Equation (2.23) gives the phase speed for internal waves when both upper and lower layers are shallow.

2.2.1. Shearing Gravitational Waves

In the above development, the basic current in both the fluid layers was assumed to be zero. Consider now an inclusion of a basic current and assume that an internal surface of density discontinuity also has a velocity discontinuity across the boundary. Let U_1 and U_2

be the equilibrium velocities in the lower and upper layer respectively. The undisturbed horizontal boundary (or the interface) will be taken at $z = 0$ and both the fluids will be assumed to be infinitely deep. The perturbation equations governing the disturbed motion are identical to the set of equations (2.13); however, they must be applied to both the layers. It is assumed that the perturbations approach to zero as z approaches $\pm\infty$. Additionally, an internal boundary condition that the pressure is continuous at the interface is applied to each layer. With a little manipulation, we obtain the following expression for the phase speed:

$$c = \frac{\rho_1 U_1 + \rho_2 U_2}{\rho_1 + \rho_2} \pm \sqrt{\frac{g}{k} \cdot \frac{\rho_1 - \rho_2}{\rho_1 + \rho_2} - \frac{\rho_1 \rho_2 (U_1 - U_2)^2}{(\rho_1 + \rho_2)^2}} \qquad (2.24)$$

A number of interesting results can be deduced from the above equation. First, when an air-sea boundary is considered, ρ_2 (for air) can be neglected in relation to ρ_1 (for water); in this case, eq. (2.24) reduces to the well-known deep-water wave formula, namely

$c = U_1 \pm \sqrt{g/k}$.

Equation (2.24) has applications in the atmospheric boundary layer where shearing-gravitational waves are often produced along an inversion surface where a density and (or) a velocity discontinuity may exist. These waves can become visible at the inversion level in the form of parallel bands of stratocumulus or altocumulus clouds known as billow clouds. The propagation of these waves in the atmospheric boundary layer has been studied by Sekera(1947), Haurwitz(1948) and Scorer(1951a) among others; the appearance of billow clouds and breaking waves at an inversion surface have been reported by Scorer (1951b), Hallet(1972) and others. As we shall discuss later in Chapter 3, an equation similar to (2.24) was used by Lord Kelvin to propose a theory of wave growth by wind.

In an oceanic environment, the magnitudes of the flow speeds U_1 and U_2 are in general small and of the same order of magnitude in the lower and the upper layer; consequently, we can assume $U_1 = U_2 = 0$. With this assumption, eq. (2.24) is reduced to,

$$c^2 = \frac{g}{k}\left(\frac{\rho_1 - \rho_2}{\rho_1 + \rho_2}\right) \qquad (2.25)$$

Equation (2.25) is similar to the deep-water phase speed equation excepting for the correction factor $\left(\frac{\rho_1 - \rho_2}{\rho_1 + \rho_2}\right)$. Across a thermocline, the difference between the density of the lower and the upper layer fluid would be about 2 kg m^{-3} giving a value of about .001 for the factor $\left(\frac{\rho_1 - \rho_2}{\rho_1 + \rho_2}\right)$. Thus the deep-water phase speed is reduced by a factor of about .03 for the same wave length; this would mean waves of long periods of oscillation. This explains why periods of oscillations of the common surface of two liquids of very nearly equal density (ex. oil on water) are very long compared to those of a free surface of similar extent. Furthermore, the potential energy of a given deformation of the common surface is diminished in the ratio $(1 - \frac{\rho_2}{\rho_1})$. As a result, waves of considerable amplitudes can be easily produced on an interface of two liquids of very nearly the same density. This fact can be used to explain the abnormal resistance experienced by ships travelling near the mouths of some of the Norwegian fjords where a layer of fresh water is often present over the salt water of the sea producing large amplitude waves at the interface (see Lamb, 1932).

2.2.2 Capillary Waves

The most ubiquitous of the ocean waves are the capillary waves which are readily produced on the ocean surface under light wind conditions; these waves are only a few centimeters in length and only a few millimeters in height and are controlled more or less by the surface tension force. A mathematical expression for the phase speed of the capillary waves can be obtained from (2.21) when we consider an air-sea interface for which $\rho_2 \ll \rho_1$; further, retaining the surface tension term we obtain

$$c^2 = \frac{\frac{\rho g}{k} + kT}{\rho \coth kh} \qquad \text{(here } \rho_1 \text{ is replaced by } \rho \text{ and } h_1 \text{ by } h)$$

Now using a deep-water approximation ($\frac{h}{L} \gg 1$, $\coth kh \sim 1$), we can simplify the above to obtain,

$$c^2 = \frac{g}{k} + \frac{kT}{\rho} = \frac{gL}{2\pi} + \frac{2\pi T}{\rho L} \qquad (2.26)$$

Equation (2.26) gives the phase speed for capillary waves. By differentiating (2.26) with respect to k, we can obtain a minimum value of c for some value of k. A little simplification gives

$$c_m^2 \text{ (minimum)} = 2g^{\frac{1}{2}} \rho^{-\frac{1}{2}} T^{\frac{1}{2}} \qquad (2.27)$$

Now divide (2.26) by c_m^2 to obtain

$$\frac{c^2}{c_m^2} = \frac{1}{2} [\frac{k_m}{k} + \frac{k}{k_m}] = \frac{1}{2} [\frac{L_m}{L} + \frac{L}{L_m}] \qquad (2.28)$$

In (2.28), k_m and L_m refer to the minimum values of the wave number and the wave length corresponding to the minimum c namely c_m.

Equation (2.28) is shown schematically in Figure 2.4. The curve in Fig. 2.4 shows the variation of the nondimensional ratio $\frac{c}{c_m}$ with respect to the variation of $\frac{k_m}{k}$ (or $\frac{L}{L_m}$). Three important regimes are shown on the graph. In the first regime, the surface tension dominates while the effect of gravity is negligible; in the second (or the intermediate) regime, neither the surface tension nor the gravity is negligible; in the third regime, the surface tension becomes negligible as gravity dominates. The point where $\frac{c}{c_m} = \frac{L}{L_m} = 1$ gives the minimum phase speed which can be found from the expression, $c_m^4 = 4gT/\rho \rightarrow c_m \sim$ 23.2 cm s^{-1} (using g = 9.8 ms^{-2} and T = .074 Nm^{-1}). From this, the values of minimum wave number and wave length can be obtained as $k_m \approx 3.6$ ra. cm^{-1}; $L_m \approx 1.7$ cm. With these values of c_m, L_m and Fig. 2.4, we can set limits for various regimes. Using a 5 percent tolerance limit, we obtain that,

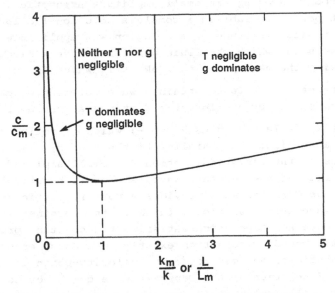

Figure 2.4: Wave speed as a function of wavelength (or wavenumber) when both gravity and surface tension are considered. (from Kinsman, 1984)

_a_. when the wave length L < 0.54 cm, effect of gravity will be less than 5 percent and the waves could be identified as capillary waves.

_b_. when the wave length L > 5.5 cm, effect of surface tension will be less than 5 percent and the waves could be identified as gravity waves.

It may be mentioned here that the discussion on capillary waves is not of direct relevance to the problem of operational wave analysis and prediction which is the main theme of this monograph; however, recent developments on satellite-sensed measurements over an ocean surface have sparked an interest in capillary waves and their role in producing backscattering of radar energy. This has led to the development of a technology capable of estimating wind speed over an ocean surface (Robinson,1985). The utility of satellite-derived wind and wave data in ocean wave analysis and modelling is discussed in Chapter 10.

2.3 Finite Amplitude Waves

The phase speed solutions derived in sections 2.1 and 2.2 were based on small-amplitude assumptions meaning that the wave amplitude is small compared to the water depth ($\frac{H}{h} \ll 1$). In reality, the waves at sea are not always of small amplitude to satisfy the above inequality. In such a situation, the phase speed solution needs to be developed without invoking the small-amplitude assumption. Gerstner was the first person to publish a complete solution (in 1802) for waves with a finite amplitude. His solution was applicable only to deep-water waves for which the fluid particles have a circular orbit. Without invoking the assumption that ak $\ll$ 1 (which is same as $\frac{\pi H}{L} \ll 1$), Gerstner was able to obtain a wave solution which describes a trochoid traced out by a point at a distance a (the amplitude) from the centre of a circle of radius $\frac{L}{2\pi}(= \frac{1}{k})$ which is rolled along the underside of a straight line parallel to the x-axis; this trochoidal motion reduces to the sinusoidal motion for small amplitude assumption. The profile of the Gerstner wave is shown schematically in Figure 2.5a. The Gerstner wave provides a more realistic description of the ocean wave than a sinusoid; further, the Gerstner wave is the only gravity wave solution which satisfies the constant-pressure, surface dynamic boundary condition exactly. The Gerstner wave is however not irrotational and can posses no velocity-potential function. The phase speed solution of the Gerstner wave comes out to be the same as that for a small-amplitude (deep-water) wave given by (2.16).

For nondivergent and irrotational fluid flow, higher order solutions for the free surface and for the wave phase speed were developed by Stokes(1847,1880). For example, a third-order solution can be developed by considering the third-order terms corresponding to the set of equations (2.9) and solving these sets of equations step-by-step. A third-order Stokes' solution in terms of the free surface displacement η and phase speed c can be expressed as;

$$\eta = \frac{1}{2}ka^2 + a \cos kx + \frac{1}{2}ka^2\cos 2kx + \frac{3}{8}k^2a^3\cos 3kx$$

$$c^2 = \frac{g}{k}(1 + a^2k^2) \hspace{3cm} (2.30)$$

Another form in which the solution is commonly expressed is given by

$$\eta = - a \cos kx + \frac{1}{2}ka^2\cos 2kx - \frac{3}{8}k^2a^3\cos 3kx \qquad (2.30)$$

The expression (2.30) for η can be derived from (2.29) by suppressing the constant $\frac{1}{2} ka^2$ and increasing kx by π.

To the first order, the Stokes' wave has the phase speed which is same as that for the small amplitude wave given by (2.16). To the third order, the Stokes' wave moves slightly faster than the small amplitude wave. Further, since $\pi(H/L) = \pi\delta$ where δ is the wave steepness, we can express Stokes' solution in terms of wave steepness. To either third or fourth order,

$H = 2a + \frac{3}{4} k^2a^3$ so that $\pi\delta = ka + \frac{3}{8} k^3a^3$. The solution to third order is,

$$\frac{\eta}{a} = - \cos kx + \frac{1}{2} \pi\delta(1 - \frac{3}{8}\pi^2\delta^2)\cos 2kx - \frac{3}{8}\pi^2\delta^2\cos 3kx$$

$$c^2 = (1 + \pi^2\delta^2)$$

The Stokes' wave as described by (2.30) is shown schematically in Figure 2.5b. Still higher order approximations to the finite amplitude waves have been worked out by Stokes and confirmed by independent

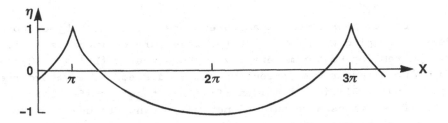

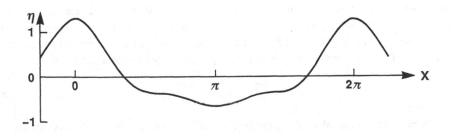

Figure 2.5:   a. The profile of a Gerstner wave b. A third-order Stokes' wave (see eq. 2.22) (from Kinsman, 1984)

calculations of Lord Rayleigh and others. For example, a fourth-order solution for the Stokes' wave is given as:

$$\frac{\eta}{a} = - \cos kx + \frac{1}{2}\pi\delta(1 + \frac{25}{24}\pi^2\delta^2)\cos 2kx$$

$$- \frac{3}{8}\pi^2\delta^2(1 - \frac{3}{4}\pi^2\delta^2)\cos 3kx + \frac{1}{3}\pi^3\delta^3\cos 4kx$$

$$c^2 = \frac{g}{k}(1 + \pi^2\delta^2 + \frac{1}{2}\pi^4\delta^4)$$

The series generated by Stokes solution has been proved mathematically to converge to a finite nonzero value.

In the above development, we have obtained a finite-amplitude wave with a rigid profile on the free surface accompanied by an irrotational motion in the fluid; this results in a slight forward movement of the fluid in the direction of the wave profile motion and a slight net transport of mass. The mathematical theory of mass transport in the direction of wave motion was first given by Stokes who obtained an expression for the wave-current speed as a function of (fluid) depth z; his expression is given by;

$$u_w^z = k^2a^2ce^{2kz} = \pi^2\delta^2ce^{2kz}$$

$$\text{at } z = o, \qquad u_w^o = \pi^2\delta^2c$$

(2.31)

As can be seen from the above expression, u_w is subject to considerable damping with depth. At the free surface (where z = o), the wave-current speed for moderately steep waves (with $\delta = 0.10$) can reach a value of about one percent of the wind speed. The wave-current speed given by (2.31) is often identified as Stokes' drift. The Stokes' drift provides a measure of net mass transport due to wave action and must be appropriately taken into account in coastal engineering problems like oil spill movement, sediment transport etc.

The concept of finite amplitude waves can be extended to capillary waves as well. Crapper (1957) obtained a novel solution to the finite amplitude capillary waves by considering a two-dimensional, non-divergent, irrotational fluid flow. An expression for the phase speed of the finite-amplitude capillary waves was obtained as

$$c = (\frac{kT}{\rho})^{\frac{1}{2}}(1 + \frac{\pi^2\delta^2}{4})^4$$

(2.32)

Comparing the above equation with (2.26), we see that with increasing value of wave steepness, the phase speed of the capillary waves decreases which is opposite of the situation with finite amplitude Stokes' waves which increase in phase speed as the wave steepness increases. Crapper further obtains a maximum value for the wave steepness δ of the finite amplitude capillary waves as 0.73, while laboratory observations (Schooley,1958) in capillary-gravity regions suggest a maximum value of δ of about 0.5; this value of maximum steepness has been used in a recent study by Donelan and Pierson(1987) to provide an interpretation of the radar backscatter signal for a satellite-based scatterometer.

2.4 Group Wave Speed and Wave Energy

The group velocity is one of the most important concepts for dispersive waves. There are several ways to define and interpret the significance of group velocity; for our purpose here, we shall consider a kinematic approach in which two wave trains of same amplitudes but slightly different wave lengths and periods are assumed to progress in the same direction. The resulting displacement can be represented as a sum of two individual disturbances η_1 and η_2, thus

$$\eta = \eta_1 + \eta_2 = a \cos(kx - \sigma t) + a \cos[(k + \Delta k)x - (\sigma + \Delta\sigma)t]$$

$$= 2a \cos \tfrac{1}{2} [(2k + \Delta k)x - (2\sigma + \Delta\sigma)t] \cos \tfrac{1}{2} (\Delta k\, x - \Delta\sigma\, t)$$

$$\approx 2a \cos(kx - \sigma t)\cos \tfrac{1}{2} (\Delta k\, x - \Delta\sigma\, t)$$

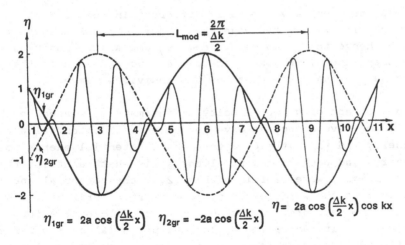

Figure 2.6: The group behaviour of an amplitude-modulated wave bounded by two conceptual waves, $\eta_1 gr$ and $\eta_2 gr$ (from Kinsman, 1984).

As can be seen from Figure 2.6, the resulting disturbance appears as a series of successive groups of waves within which the amplitude varies from 0 to 2a, the modulation term being $\cos\tfrac{1}{2}(\Delta k\, x - \Delta\sigma\, t)$. The distance between centres of successive groups is given by

$\frac{2\pi}{\Delta k}$ and the time required for the disturbance to travel this distance
is $\frac{2\pi}{\Delta\sigma}$. The speed of the disturbance as a group wave is then obtained
as

$$c_g = \frac{\Delta\sigma}{\Delta k} \qquad \text{which in the limit becomes}$$

$$c_g = \frac{d\sigma}{dk} = \frac{d}{dk}(kc) = c + k\frac{dc}{dk} = c - L\frac{dc}{dL} \qquad (2.33)$$

Equation (2.33) is the formula for group wave speed or group
velocity as it is commonly known. Re-writing the dispersion relationship (2.10) as $\sigma^2 = gk \tanh kh$ and taking derivatives, we obtain,

$2\sigma\frac{d\sigma}{dk} = g(\tanh kh + kh \, \text{sech}^2kh)$; this can be further simplified to

write $\frac{d\sigma}{dk} = \frac{1}{2}c(1 + \frac{2kh}{\sinh 2kh})$ $\qquad (2.34)$

Using the deep- and shallow-water approximations in (2.34), we
obtain,

$$c_g = \frac{c}{2} \qquad \text{for deep-water waves}$$
$$(2.35)$$
$$c_g = c \qquad \text{for shallow-water waves}$$

Thus the group wave speed is different in deep and shallow
waters. In shallow water, the wave phase speed depends only on the
water depth, hence the group wave speed is same as the individual wave
speed. In deep waters, the waves are dispersive and as such the group
wave speed is only half the speed of the individual wave.

The concept of group velocity is intimately related to the
transmission of wave energy. A wave possesses both potential and
kinetic energy and the rythmic conversion of potential energy to
kinetic energy and back to potential energy is what maintains the
oscillations. For a wave profile of Fig. 2.1, the potential energy
can be defined as the work necessary to distort a horizontal sea
surface into the wave profile. Using a wave profile of the form
$\eta = a \cos(kx-\sigma t)$, it can be shown that the potential energy for such a
wave profile over one wave length L is given by the integral,

$$E_p = \int_o^L \frac{\rho g}{2}\eta^2 dx \qquad (2.36)$$

In (2.36), ρ is the water density, g is gravity and dx is an
increment along the wave length in the x direction in which the wave
is progressing. If we consider a unit width for the wave profile,
the horizontal area of the sea surface over one wave length will be

L · 1 = L, hence the average potential energy per unit horizontal surface area is given by,

$$E_p = \frac{\rho g}{2L} \int_o^L \eta^2 \, dx \qquad (2.37)$$

In (2.37), the water density ρ is assumed constant, and the surface tension effects are neglected. If we substitute for η a progressive wave form as given earlier (eq. 2.1), we obtain on integration

$$E_p = \frac{1}{4} \rho g a^2 \qquad (2.38)$$

Next, the average kinetic energy per unit surface area over one wave length can be written as;

$$E_k = \frac{\rho}{2L} \int_o^L \int_{-h}^{\eta} (u^2 + w^2) \, dx dz \qquad (2.39)$$

In (2.39), u and w are the horizontal and the vertical velocity components, h is the water depth measured (downward) from the undisturbed water surface and ρ is the water density, assumed constant. Now, substitute for u and w in terms of their solutions applicable to deep-water, namely,

$$u = a\sigma e^{kz} \cos(kx - \sigma t)$$
$$w = a\sigma e^{kz} \sin(kx - \sigma t) \qquad (2.40)$$

Simplify using the limit of deep-water (h → ∞) and obtain

$$E_k = \frac{\rho g a^2}{4} e^{2k\eta}$$

This can be further simplified by using small amplitude assumption, namely

$2k\eta \ll 1$ hence $e^{2k\eta} \to 1$.   With this, we

have $\qquad\qquad E_k = \frac{\rho g a^2}{4} \qquad (2.41)$

Comparing (2.38) and (2.41), we see that to the first order of approximation, there is an equipartition of energy between the potential and the kinetic energy of a sinusoidal progressive wave. Further, we can write the total energy of a sinusoidal wave per unit area as $E = E_p + E_k = \frac{1}{2} \rho g a^2$. In terms of wave height H, this can be written as

$$E = \frac{1}{8} \rho g H^2 \qquad (2.42)$$

The wave energy per unit area is equal to that necessary to raise a layer of water of thickness equal to the wave height H through a distance $\frac{H}{8}$. Further, it can be shown (see Kinsman,1984) that the mean wave power (rate of transmission of energy) per unit crest width averaged over a wave period is given by,

$$\text{Power} \quad P \; = \; E \cdot \frac{c}{2}(1 + \frac{2kh}{\sinh 2kh})$$

using (2.34),

$$P \; = \; E \cdot c_g$$

Once again, using the deep- and shallow-water approximations we obtain

$$
\begin{aligned}
P &= E \cdot \frac{c}{2} \quad &\text{for deep-water waves} \\
P &= E \cdot c \quad &\text{for shallow-water waves}
\end{aligned}
$$

(2.43)

We see from these equations that in deep water, wave energy is transmitted at half the wave phase speed which is the group velocity for the deep-water waves; this has the effect that, where a wave train of fixed period travels out of the generating area, the leading wave dies out quickly and the front of the wave train progresses not at the speed of the individual wave but at an appropriate group wave speed. As we shall see later, wave prediction models based on the spectral energy balance equation include a term which uses the group velocity c_g for transmission of wave energy.

REFERENCES

Crapper, G.D., 1957: An exact solution for progressive capillary waves of arbitrary amplitude. J. Fluid Mechanics, 2, 532-540.

Donelan, M.A. and W.J. Pierson, 1987: Radar scattering and equilibrium range in wind-generated waves with application to scatterometry. J. Geophysical Res., 92, C5, 4971-5029.

Hallet, J., 1972: Breaking waves at an inversion. Monthly Weather Review, 100, 133-135.

Haltiner G.J. and R.T. Williams, 1980: Numerical prediction and dynamic meteorology; second edition. John Wiley and Sons, 477 pp.

Haurwitz, B., 1947: Uber Wallenbewgungen an der Grezflache Zweir Luftschichten mit linearen Temperaturgefalle. Beitr. Physik fr. Atmos., 19, 47-54.

Lamb, H., 1932: Hydrodynamics. Cambridge University Press, U.K., 738 pp., 371-372.

Kinsman, B., 1984: Loc. cit. (Ch. 1)

Neumann, G. and W.J. Pierson, 1966: Principles of Physical Oceanography. Prentice-Hall, U.S.A., 545 pp., 272-275.

Osborne A.R. and T.L. Burch, 1980: Internal solitons in the Andaman
Sea. Science, 208, No. 4443, 451-460.

Pond, S. and G.L. Pickard, 1983: Introductory Dynamical Oceanography.
Pergamon, U.K., 329 pp., 278-279.

Robinson, I.S., 1985: Satellite Oceanography: An introduction for
oceanographers and remote-sensing scientists. Ellis Horwood Ltd.,
U.K., 455 pp.

Schooley, A.H., 1958: Profiles of wind-created gravity waves in the
capillary-gravity transition region. J. Marine Res., 16, 100-108.

Scorer, R.S., 1951a: On the stability of stably stratified shearing
layers. Q.J. Royal Met. Soc., 77, 76-84.

Scorer, R.S., 1951b: Billow clouds. Q. J. Royal Met. Society, Vol. 77,
235-240.

Sekera, Z., 1948: Helmholtz waves in a linear temperature field with
vertical wind shear. J. Meteorology, Vol. 5, 93-102.

Stokes, G.G., 1847: On the theory of oscillatory waves. Trans.
Cambridge Phil. Soc., U.K., Vol. 8, 441 pp.

Stokes, G.G., 1880: Supplement to a paper on the theory of oscillatory
waves. Mathematical and Physical papers, Vol. 1, Cambridge University
Press, U.K., 314-326.

CHAPTER 3
WAVE GENERATION, PROPAGATION AND DISSIPATION

3.1 Historical Notes

The problem of wave generation by wind has attracted the attention of naturalists, philosophers and scientists alike, for several centuries. In the historical past, the Greek philosopher Aristotle (384-322 B.C.) recognized the importance of wind on wave generation while the Italian naturalist and historian Pliny (Pliny the Elder, 23-79 A.D.) observed the sobering effect of an oil layer on the surface water waves. The versatile Leonardo da Vinci made some interesting observations of wave fields and wave groups around 1500 while the great mathematician Euler proposed in 1755 that the best way to obtain the general solution of the fluid motion and in particular of wave motion was to consider several special cases and build a suitable basis for further development. The American statesman and scientific explorer, Benjamin Franklin made laboratory observations (in 1762) of waves of long periods of oscillations at the interface of oil and water; later (in 1774) Franklin and his co-workers while making observations of wind waves recorded that "Air in motion, which is wind, in passing over the smooth surface of the water, may rub, as it were, upon the surface and raise it into wrinkles which, if wind continues are the elements of future waves." Franklin and his co-workers had correctly perceived the wrinkles (or the capillary waves) as the forerunners of the wind-induced gravity waves.

3.2 Early Theories of Kelvin, Helmholtz and Jeffreys

The classical Kelvin-Helmholtz instability theory may be considered as among the earliest of the wave generation theories based on a mathematical argument. The theory is based primarily on a paper by Lord Kelvin (Sir W. Thompson) entitled 'hydrokinetic solutions and observations (1871)[1]' and another paper by Helmholtz entitled 'Uber discontinusrlich Flüssigkeitsbewegungen' (on discontinuous fluid motion, 1868)[2]. The theory assumes air pressure to be 180° out of phase with the (water) surface elevation and for sufficiently large

---

[1] Nature, Vol. 1, 1871
[2] Philosophical Magazine, London, U.K., November 1868

wind speed, the pressure distribution over the surface due to the
Bernoulli effect can cause an infinitesimal-amplitude sinusoidal wave
to grow against the restoring forces of gravity and surface tension.
The paper by Kelvin developed an equation similar to (2.24) and using
a stability analysis obtained a minimum wind speed of 6.5 m s^{-1} necessary for waves to grow; this minimum is the absolute minimum for waves
travelling in the same direction as the wind. Most field observations,
however, suggest that waves on water can develop and grow at wind
speeds much lower than the predicted (absolute) minimum of 6.5 m s^{-1}
required by the Kelvin-Helmholtz instability theory.

The next major theory on wave generation was put forward by
Jeffreys (1924, 1925) and is known as the 'sheltering hypothesis'.
Jeffreys considered a wave profile as shown in Figure 3.1 with normal
(R_N) and tangential (R_T) components of rates of energy transfer
from wind to wave.

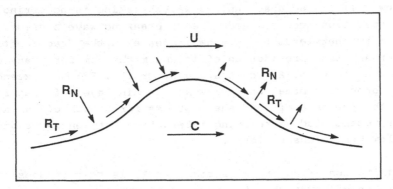

Figure 3.1: A water wave with tangential (R_T) and normal (R_N)
components of rates of energy transfer from wind to wave.

Jeffreys neglected the tangential wind stress and assumed that
the only important mechanism for transferring energy from wind to
water was the difference of normal pressure between the windward and
the leeward side of a wave crest. He further hypothesized that the
waves would continue to grow only if the normal energy flux integrated
over the water surface (of the wave crest) exceeds the rate at which
the energy is dissipated by molecular viscosity. Mathematically, this
can be expressed as,

$$s\rho_a(U-c)^2\, c > 4\mu g \qquad\qquad (3.1)$$

Here U is the wind speed, c is the wave phase speed, ρ_a is the air
density, μ is the coefficient of viscosity for water and s is the
dimensionless constant called a sheltering coefficient. Jeffreys hypothesized that a turbulent wind blowing over a pre-existing wave crest

would act like air blowing past a blunt body causing a 'boundary layer separation' on the downwind side of the wave crest; the boundary layer would presumably re-attach itself on the upwind side of the next wave crest. This hypothetical flow pattern, according to Jeffreys would cause a pressure asymmetry resulting in wave growth. However, laboratory experiments conducted by other investigators over solid waves in wind tunnels indicated that pressure differences on two sides of the wave was too small for Jeffreys' mechanism to be effective for wave growth. This led Ursell(1956) to conclude that the hypothesis of sheltering in its simplest form is insufficient to explain the growth of waves and mechanisms involving a drag must be considered.

A suggestion by Barnett and Kenyon(1975) that Jeffreys' hypothesis of boundary layer separation may yet emerge to be important for wave growth has prompted Banner and Melville(1976) to examine this hypothesis closely in a laboratory study. Banner and Melville found that air flow will not separate unless wave breaking is occurring, in which case separation occurs ahead of each breaking wave's crest. The wave breaking further leads to a large increase in the drag coefficient for air and thus provides an efficient mechanism for transfer of energy from wind to wave. A recent study by Croft(1985) has extended this idea to propose a model which invokes a thin layer of (water) droplets in the presence of wind shear stress for growth of surface waves. Other recent studies relating interaction between waves and turbulent flow are discussed later.

A limitation of Jeffreys' formula (3.1) is that it does not allow the growth of waves when the wave phase speed (c) is greater than the wind speed (U). Sverdrup and Munk(1947) in their pioneering work on wave prediction included the tangential wind stress $\tau = \rho_a C_D U^2$ (where C_D is the drag coefficient for air) in the wave generation process. The average rate at which energy is transmitted to waves by the tangential component of the wind stress may be expressed as,

$$R_T = \frac{1}{L} \int_0^L \tau u \, dx \qquad (3.2)$$

Here u is the horizontal component of the fluid particle velocity at the sea surface and L is the wave length over which the average rate of energy input is calculated. For a finite amplitude wave, u is the surface value of the Stokes' drift given by (2.31) as $u = \pi^2 \delta^2 c$. Substituting this value in (3.2) and simplifying, the criterion for wave growth can be rewritten as,

$$2\rho_a C_D U^2 c \pm s\rho_a (U - c)^2 c > 4\mu g \qquad (3.3)$$

Eq. (3.3) allows the growth of waves even when the phase speed c is greater than the wind speed U; this is more realistic and has been confirmed by observations. As we shall see in the following Chapter, Sverdrup and Munk utilized in their wave prediction technique, the physical concept embedded in eq. (3.3).

3.2 Wave Generation and Growth by Linear Processes

The publication of Ursell's(1956) review paper on 'wave generation by wind' appears to have sparked two major independent and complementary theories of wave generation; these are now well recognized as Phillips'(1957) linear growth mechanism and Miles'(1957) exponential growth mechanism. Phillips theory was an improvement over an earlier wave generation theory of Eckart(1953) in which wind was represented by a random distribution of normal pressure in the form of idealized circular gusts over a finite storm area. Phillips(1957) proposed that waves grow by a resonance mechanism when the speed and the length of the atmospheric fluctuations match those of the water waves; he further proposed that the waves continue to grow by this mechanism until the wave slopes become large enough that nonlinearities which are neglected in his theory become important.

Phillips assumed that the fluctuating pressure upon the water surface was responsible for the birth and early growth of waves; he further allowed the pressure field to evolve in a random way to be convected over the water surface by the mean wind. Phillips' theory could be divided into two parts by considering the time from the onset of a turbulent wind to be either much less than (initial stage) or much greater than (principal stage) the time scale for the development of the pressure fluctuations. The major growth of the gravity waves, in Phillips' theory, takes place in the principal stage of the development where the energy grows linearly with time as dictated by the equation,

$$\frac{d}{dt}E(k) = \frac{\pi \sigma^2}{2 \rho_w g} F(k, -\sigma) \tag{3.4}$$

Here $E(k)$ is the wave energy expressed as a function of wave number k, σ is the wave angular frequency, ρ_w is the water density and $F(k, -\sigma)$ is the three-dimensional spectrum of pressure fluctuations. The function $F(k, -\sigma)$ can be specified based on correlation studies on atmospheric pressure fluctuations and the turbulent boundary layer. A brief analysis on prescribing a suitable expression for the pressure spectrum has been given by Barnett(1968).

Phillips' resonance theory predicts a minimum wind speed of 23 cm s^{-1} to generate capillary waves of wave lengths 1.7 cm (The mean wind speed is to be measured at a height of about one wave length above the water surface). Field studies conducted by Roll(1951) indicate waves with wave lengths of about 1.7 cm developed under light wind conditions of 1 to 4 m s^{-1}, thus supporting Phillips' theory. However, the resonance theory is not considered to be the mechanism to explain major growth of wind waves because the observed pressure fluctuations are too small and further, the observed energy growth rates are more nearly exponential than linear. In summary, Phillips' resonance mechanism appears important in bringing the energy level of waves from zero to a point where other mechanisms such as exponential growth or instability can take over.

Miles(1957,1959) considered the generation of water waves due to shear flow instability in the coupled air-water system. Miles attempted to improve upon the classical Kelvin-Helmholtz instability theory by considering a wind profile continuously increasing with height and a pre-existing water wave which induces a disturbance in the shear flow; that part of the induced pressure disturbance which is in phase with the wave slope does the work and causes it to grow. This coupled mechanism results in an exponential growth rate for the wave energy; this growth rate in Miles' theory is obtained in terms of derivatives of the mean wind profile evaluated at the critical layer where the mean wind speed (U) and the phase speed (c) of the water wave are equal. An expression for the rate of energy transfer is given by,

$$\frac{d}{dt}E(k) \;=\; -\;\frac{\pi \sigma}{2g}\cdot\frac{\rho_a}{\rho_w}\;\left(\frac{d^2\bar{U}}{dz^2}\bigg/\frac{d\bar{U}}{dz}\right)_{cr}\;|w|^2 E(k) \tag{3.5}$$

Here $\bar{U}$ is the mean wind speed, ρ_a is the air density, ρ_w the water density and w is the vertical velocity of the water surface in response to a periodic unit amplitude surface displacement of the phase velocity c. The derivatives $\frac{d\bar{U}}{dz}$ and $\frac{d^2\bar{U}}{dz^2}$ are to be evaluated at the critical layer where the wind speed equals the phase speed of the water waves.

Equation (3.5) provides positive wave growth rates in general, since a normal wind profile has negative curvature and positive slope. For a logarithmic wind variation (commonly found under neutral atmospheric stability) in the atmospheric boundary layer, the energy transfer increases with decreasing height of the critical layer. As the waves increase in wave length, the phase speed increases and so does the height of the critical layer; this in turn reduces the wave growth rate as given by (3.5). On the other hand, if the critical layer lies very close to the water surface in a laminar sublayer with a linear velocity profile, then the energy transfer would vanish according to (3.5). Thus Miles' mechanism is particularly effective for waves with phase velocities appreciably lower than the wind speeds. For waves with phase velocity greater than the maximum wind speed or for waves travelling at angles greater than 90° to the wind direction, Miles' theory does not provide any growth nor any decay.

Observations of wave growth were made by Snyder and Cox(1966) during a field experiment in which the development of a single spectral component was determined by towing a four-buoy array downwind at a constant speed in the lee of the Eleuthera island in the Bahamas. The wave array was tuned to waves having a wave length of 17 m. The growth of the wave component was determined from the analysis of the wave record while the mean wind speed was determined by an aerovane mounted at 6.1 m above the water surface. The wave growth data was analyzed using a Miles-Phillips growth formula given by,

$$\frac{d}{dt} E(t) = \alpha + \beta E(t) \tag{3.6}$$

Each set of data were analyzed assuming constant values of α and β and using a least square technique to minimize the variance between observed and computed values. The best-fit values of the parameters (α and β) were plotted against the mean wind speed corresponding to each dataset. Figure 3.2 shows the scatter diagram of points showing variation of β and U the mean wind speed. The data points were divided into two catagories, one in which the E (energy) values were less than 50 cm²s for the duration of a dataset and the other in which the E values varied arbitrarily. These two catagories of data are shown by using the symbols ◻ and ■ in Fig. 3.2. The Fig. also shows the predicted growth curves arising from the theories of Jeffreys and Miles (labelled J and M) and a straight line obtained by fitting an equation $\beta = \frac{\rho_a}{\rho_w} (U_c - \sigma)$ to the data points; here U_c is the mean wind speed at a critical wave length above the water surface, σ is the wave frequency and $\frac{\rho_a}{\rho_w}$ is the ratio of the density of air to that of water. The wind speed U_c is normalized to U via a logarithmic wind profile. The Fig. shows clearly that the growth curves theorized by Jeffreys and Miles are almost an order of magnitude too small when compared with the observed growth rates. A second set of data collected by Barnett and Wilkerson (1967) showed once again that Miles' theory underpredicted the wave growth by almost an order of magnitude.

The enhanced observed wave growth as revealed by the above studies sparked a number of theoretical and observational studies in the last twenty years. In the original studies of Miles(1957, 1959),

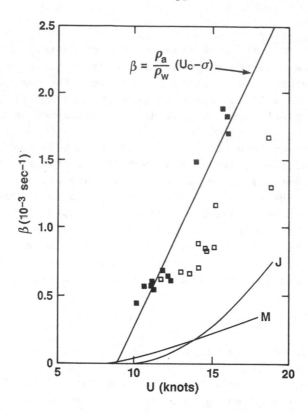

Figure 3.2: Scatter diagram showing variation of β with U the mean wind speed. Curves J and M denote growth curves theorized by Jeffreys and Miles respectively, while the straight line is based on the equation as shown (from Snyder and Cox, 1966).

turbulence in the air flow over the water surface was invoked to provide a logarithmic wind profile but was otherwise ignored. In his later study, Miles(1967) attempted to modify his theory to include interaction between waves and the air turbulence. Following Miles' suggestion, Davies(1969, 1972) investigated the nature of turbulent air flow over a wavy boundary and how the air flow could do work on the waves. Subsequent studies (Gent and Taylor, 1976; Riley, Donelan and Hui, 1982, Al-Zanaidi and Hui, 1984) have considered the impact of turbulence by using eddy viscosity models of varying degrees of complexity. A recent study by Jacobs(1987) reviews some of the earlier studies and presents an asymptotic theory for turbulent flow over water waves in which the eddy viscosity is assumed to vary linearly with distance from the water surface. Jacobs obtains growth rates which compare favourably with observations for the most rapidly amplifying waves; Jacobs' study further demonstrates that waves travelling against the wind, or more rapidly than the wind, transfer energy and momentum from water to air. However, Jacobs' model as well as other

turbulence models obtain growth rates which are lower than those measured by Snyder and Cox(1967) and others.

A number of laboratory and field studies on wave growth and
over-water pressure measurements have been reported in the last twenty
years. Shemdin and Hsu(1967) obtained laboratory measurements of aero-
dynamic pressure distribution at the interface between air and simple
progressive water waves, while field measurements of pressure fluctuations have been made and reported by Dobson(1971), Elliot(1972) and
Synder(1974) and more recently by Snyder et al (1981) among others.
Some of the earlier studies (ex. Dobson, Elliot) show considerable
disagreement in the various sets of data collected during these field
measurements by these investigations; part of this disagreement could·
be attributed to the demanding nature of the experimental set up and
related instrumentation. The study conducted by Snyder et al(1981) was
designed to resolve the disagreement found by earlier investigators;
their measurements together with a reexamination of earlier results
have been summarized in the form of an expression for the growth
factor given by,

$$
\beta \begin{cases} = [0.2 \text{ to } 0.3] \left\{ \dfrac{U_s \cos(\emptyset - \theta)}{c} - 1 \right\} \dfrac{\rho_a}{\rho_w} & \text{for} \quad 1 < \dfrac{U_s \cos(\emptyset - \theta)}{c} < 3 \\[3em] = 0 & \text{for} \quad \dfrac{U_s \cos(\emptyset - \theta)}{c} \leq 1 \end{cases}
$$

(3.7)

In (3.7), U_s is the mean wind speed at a height of 5 m above the water
surface in the atmospheric boundary layer, $\emptyset$ is the wind direction,
θ is the direction of the wave component and ρ_a and ρ_w are defined
earlier. Equations (3.6) and (3.7) describe the exponential growth of
a wave component only when the component of wind speed in the direction of the wave component exceeds the phase speed of the wave component; the rate of growth increases as the deficit of the wind speed
component over the phase speed of the wave component increases. More
recent studies (Plant,1982; Mitsuasu and Honda, 1982) based on laboratory data as well as on field measurements have obtained the growth
factor β in terms of U_* the friction velocity, an important atmospheric boundary layer parameter defined by,

$$
U_* = \sqrt{\tau/\rho_a} = U \sqrt{C_D}
$$

(3.8)

In (3.8), τ is the wind stress, ρ_a is the air density and C_D is the
drag coefficient for air. The Jacobs' model (mentioned earlier)
obtains the wave growth which compares favourably with Plant's (1982)

study but is considerably lower than the growth rates obtained by
Mitsuasu and Honda(1982) which are based on wave-tank experiments.

In summary, most observational studies on wave generation
reveal the wave growth rates to be considerably higher than those
predicted by Miles' original studies (1957,1959) as well as subsequent
other studies which have attempted to improve upon Miles' growth
mechanism. The present status of these studies could be aptly summarized in a comment by Riley, Donelan and Hui(1982) namely, "The extension of Miles' theory moves us further in the direction of believing
that theories of this type are at most only part of the explanation
for wave amplification or decay by wind".

3.3 Wave Growth by Non-linear Interactions

As the surface gravity waves continue to grow in an active
wind field, the average wave slope continues to increase. The average
wave slope is a measure of the nonlinearity of the waves and as the
waves continue to grow the nonlinear terms of the governing equations
become important enough that the linear theories of Miles and Phillips
no longer apply. Research and developmental work on the nonlinear wind
wave problem expanded following an important paper by Phillips (1960)
which showed that under certain conditions, the third-order perturbations can become large and unsteady. The conditions for unstable
perturbations to exist are called the resonance conditions and are
defined by the equations,

$$k_1 \pm k_2 \pm k_3 \pm k_4 = 0$$
$$\sigma_1 \pm \sigma_2 \pm \sigma_3 \pm \sigma_4 = 0$$

(3.9)

Here the wave number and the frequency pairs (k_i, σ_i) are those of
free primary waves which individually satisfy the wave phase speed
relation (2.10) which can be re-written as $\sigma^2 = gk \tanh kh$. Phillips
(1960) showed that there exists a configuration of three wave numbers
k_1, k_2 and k_3 which interact to give a continuous transfer of energy
to a distinct fourth wave with wave number k_4 whose amplitude grows
linearly with time. Phillips' work was extended by theoretical work of
Longuet-Higgins(1962), Benny(1962) and others. Longuet-Higgins(1962)
calculated the interactions for the simplest possible case, namely
when two of the three primary wave numbers are equal (i.e. $k_1 = k_3$),
while Benny(1962) derived the complete set of interaction equations
governing the time dependence of the resonant modes. The effect of
resonant interactions on the entire energy spectrum of the wind waves
was derived by Hasselmann(1962,1963), who carried out the perturbation
analysis to fifth order using a generalized technique applicable to
any system involving weak interactions between ocean wave fields or

other random fields (ex. scattering of light and sound in the atmosphere). Hasselmann developed the energy transfer equations in terms of
coupling coefficients characterizing the interactions between the wave
field and its physical environment. A typical energy transfer equation
developed by Hasselmann(1962) can be expressed in a simplified form
as,

$$\frac{\partial}{\partial t} E(k_4) = \int \int \int_{-\infty}^{\infty} \frac{9\ \pi\ g^2 \sigma_4 D_4}{4\ \rho_w^2\ \sigma_1^2 \sigma_2^2 \sigma_3^2 \sigma_4^2} \{D_4 \sigma_4 E_1 E_2 E_3\ +\ D_3 \sigma_3 E_1 E_2 E_4$$

$$-\ D_2 \sigma_2 E_1 E_3 E_4\ -\ D_1 \sigma_1 E_2 E_3 E_4\} \ \delta(\sigma_1\ +\ \sigma_2\ -\ \sigma_3\ -\ \sigma_4)\ \cdot$$

$$\delta(k_1\ +\ k_2\ -\ k_3\ -\ k_4)\ dk_1 dk_2 dk_3 \qquad\qquad (3.10)$$

In (3.10), $E_j = E(k_j)$ is the energy of the wave train with wave number
k_j, D_j are the complicated coupling coefficients that are functions of
wave numbers k_1 to k_4 and $\delta(\ )$ are the delta functions.

Eq. (3.10) describes the interactions between three active wave components, which determine the interaction rate, and a passive fourth
component k_4, which receives energy from the first three components
(k_1, k_2, k_3) but has no direct influence on the interaction. The energy
transfer equation (3.10) and other similar equations conserve energy
over the entire spectrum. These transfer equations are derived assuming the Boltzmann hypothesis of statistical independence of interacting particles and hence the integrals on the right side of (3.10)
are often identified as Boltzmann transfer-integrals or simply as
Boltzmann integrals.

The resonant interactions by themselves cannot be considered as a
source for wave generation or wave dissipation, since they do not
change the total wave energy. However, these resonant interactions may
play an important role in wave growth by distributing to low frequencies the wave energy supplied by the wind to the high frequency
portion of the wave spectrum. During the JONSWAP (Joint North Sea WAve
Project; see Hasselmann et al, 1973), field experiment, extensive
measurements of wave growth under well-defined wind conditions were
made and it was shown that the rapid growth of low frequency waves was
primarily associated with nonlinear energy flux due to resonant wave-
wave interaction; further, the nonlinear energy transfer produced a
pronounced (energy) peak which migrated towards the low frequency
waves. These important discoveries have led to a class of wave models
known as parametric wave models in which the shape of the wave spectrum is defined by a small number of parameters whose growth rates can
be calculated provided the wind forcing can be prescribed appropriately. Recent studies by Hasselmann and Hasselmann(1985) and Hasselmann

et al (1985) have demonstrated that the Boltzmann integrals of eq.
(3.10) can be evaluated completely using an efficient integration
algorithm which can be handled by present generation supercomputers;
this has led to the development of the third generation operational
wave models. More details of the parametric and the third generation
wave models are presented in Chapter 5.

3.4 Wave Propagation and Dissipation

3.4.1 Wave Propagation

After being generated (say by a mid-latitude weather disturbance or a tropical storm), wind waves can propagate for great distances over the oceans, their travel being largely uninterrupted until
they break and dissipate upon reaching a coast. Observational studies
by Barber and Ursell(1948) and Munk et al (1963) have shown that wind-
waves generated in a storm can travel as far away as halfway around
the world with very little attenuation. Munk et al have recorded
swells off the California coast which were generated by storms near
Australia, New Zealand and Antarctica. According to Phillips(1959),
swells suffer very little molecular viscous dissipation and the level
of turbulence in the ocean does not affect their propagation significantly; further, the waves do not seem to be affected by propagation
through zones of steady wind such as the Trade Wind Belt. The slow
decay of swells outside the generating area is due to the fact that
waves are nearly irrotational and nearly linear and therefore are not
dissipated by either turbulent friction or molecular viscosity. For
propagation over long distances, the waves seem to obey the linear
theory of wave propagation from a limited initial disturbance; this
theory leads to the concept of group velocity as the signal velocity
of the wave energy (see section 2.4). In practice, the concept of
group velocity can be used to predict the arrival time of waves at a
coastal location if the position of the storm over the ocean and the
time of its occurrence are known.

In operational numerical models, the wave propagation is
generally computed by the term $c_g.\nabla E$ where c_g is the group
velocity and ∇E is the gradient of energy for the waves. If a wave
model operates over deep waters only, c_g will refer to the deep-
water group velocity and the term $c_g.\nabla E$ can be calculated for
individual wave numbers. The propagation of wave energy for different
wave numbers is accomplished along great circle arcs along which the
potential energy is conserved. For wave models operating in shallow
waters, wave refraction will have to be included in the wave propagation algorithm; more details of the wave propagation schemes in operational models are presented in Chapters 5 and 6.

3.4.2 Wave Dissipation

Dissipation of ocean wind waves is thought to occur primarily
when waves break during the wave generation in a storm and later when
the waves break as they approach a shoreline. Near a shoreline with
evenly sloping beaches, wave breaking can occur in two ways: wave
plunging and wave spilling. The plunging breaker has a well rounded
back and a concave front and can form with a wave steepness $\delta (=\frac{H}{L})$ of
about 0.005 and an offshore wind. Spilling breakers are concave on
both faces and are produced by waves with steepness greater than 0.01.
According to Stokes(1880), a wave breaks when water particles in its
crest advance faster than its profile and this happens when the crest
angle exceeds 120°. It can be shown that this limiting value of the
crest angle corresponds to a wave steepness of 1/7, a value which
Stokes had theorized for maximum wave steepness. Most observations of
wave steepness give a value of δ from 0.1 to 0.008. Away from the
seashore, wave breaking is generally visible in the form of white-
topped waves commonly known as whitecaps or whitehorses. According to
Phillips(1963), wave breaking over the open sea can occur by two processes: one, in which the low frequency waves intersect (or overtake)
each other and the other in which short wind-generated waves are over-
taken by longer waves; Phillips further proposes a mechanism by which
short wind-generated waves extract energy from the long waves through
a 'radiation stress'; this would always lead to the attenuation of
waves as theorized by Hasselmann(1971).

Over an open ocean, wave breaking and the consequent whitecapping is believed to be the most dominant dissipative mechanism.
Wave breaking represents a localized, strongly nonlinear interaction
process which cannot be treated by the standard perturbation techniques applicable to weak interactions as discussed in section 3.3. In
recent years, a number of important studies have been reported on wave
breaking, wave dissipation and spectral characteristics, notable among
them are those by Banner and Phillips(1974), Hasselmann(1974) and
Phillips(1985). Hasselmann(1974) investigated the effect of whitecapping on the spectral energy balance by expressing the whitecap interactions in terms of an equivalent ensemble of random pressure pulses.
He further derived a damping coefficient for whitecapping which was
shown to be proportional to the wave angular frequency σ. The dissipation coefficient and the associated dissipation function were found
to be consistent with the structure of the energy balance derived
from the JONSWAP data. Phillips(1985) in his recent study expanded
Hasselmann's approach and obtained an estimate for the average rate of
spectral energy loss resulting from wave breaking. Phillips further
postulated that the process of energy input from the wind, loss by

wave breaking and the net transfer by nonlinear interactions are of comparable importance throughout the equilibrium range of wind-generated waves. With this hypothesis, Phillips obtained a frequency spectrum which can be expressed as proportional to $U_* g \sigma^{-4}$ where U_* is the friction velocity; this form is similar to the one obtained empirically by Toba(1973).

In shallow waters, wave dissipation occurs due to a number of processes like wave breaking, wave interaction with the bottom causing wave refraction, wave shoaling etc. and wave attenuation due to soft bottom particularly in the vicinity of river deltas. Several studies have been reported on each of the above mentioned processes. A detailed discussion on some of these studies will be presented in Chapter Six.

REFERENCES

Al-Zanaidi, M.A. and W.H. Hui, 1984: Turbulent air flow over water waves - a numerical study. J. Fluid Mechanics, 148, 225-246.

Banner, M.L. and W.K. Melville, 1976: On the separation of air flow over water waves. J. Fluid Mechanics, 77, 825-842.

Banner, M.L. and O.M. Phillips, 1974: On the incipient breaking of small scale waves. J. Fluid Mechanics, 65, 647-657.

Barber, N.F. and F. Ursell, 1984: The generation and propagation of ocean waves and swell. I. Wave periods and velocities. Phil. Trans. Royal Soc., U.K., Series A, 240, 527-560.

Barnett, T.P., 1968: On the generation, dissipation and prediction of ocean wind waves. J. Geophysical Research, 73, 513-529.

Barnett, T.P. and K.E. Kenyon, 1975: Recent advances in the study of wind waves. Report on Progress in Physics, 38, 667-729.

Barnett, T.P. and J.C. Wilkerson, 1967: On the generation of wind waves as inferred from air-borne measurements of fetch-limited spectra. J. Marine Research, 67, 3095-3102.

Benny, D.J., 1962: Non-linear gravity wave interactions. J. Fluid Mechanics, 14, 577-584.

Croft, A.J., 1985: A model for surface wave growth. Offshore and coastal Modelling, Eds. Moscardini and Robson, Springer-Verlag, Lecture notes on coastal and estuarine studies, 12, 165-185.

Davis, R.E., 1969: On the high Reynolds number flow over a wavy boundary. J. Fluid Mechanics, 36, 337-346.

Davis, R.E., 1972: On prediction of the turbulent flow over a wavy boundary. J. Fluid Mechanics, 52, 287-306.

Dobson, F.W., 1971: Measurements of atmospheric pressure on wind-generated sea waves. J. Fluid Mechanics, 48, 91-127.

Eckart, C., 1953: The generation of wind waves over a water surface. J. Applied Physics, 24, 1485-94.

Elliot, J.A., 1971: Microscale pressure fluctuations measured within the lower atmospheric boundary layer. J. Fluid Mechanics, 53, 351-384.

Gent, P.R. and P.A. Taylor, 1976: A numerical model of the air flow above water waves. J. Fluid Mechanics, 77, 105-128.

Hasselmann, K., 1962: On the nonlinear energy transfer in a gravity wave spectrum - Part I. J. Fluid Mechanics, 12, 481-500.

Hasselmann, K., 1963: On the nonlinear energy transfer in a gravity wave spectrum - Part 3. J. Fluid Mechanics, 15, 385-395.

Hasselmann, K., 1971: On the mass and momentum transfer between short gravity waves and large-scale motions. J. Fluid Mechanics, 50, 189-205.

Hasselmann, K., 1974: On the spectral dissipation of ocean waves due to whitecapping. Boundary-Layer Meteorology, 6, 107-127.

Hasselmann, K., et al., 1973: Measurements of wind-wave growth and swell decay during the joint North Sea Wave Project (JONSWAP). Duet. Hydrog. Zeit., Reiche A12, 95 pp.

Hasselmann, S., and K. Hasselmann, 1985: Computations and parameterization of the nonlinear energy transfer in a gravity-wave spectrum - Part 1. J. Physical Oceanography, 15, 1369-1377.

Hasselmann, S., K. Hasselmann, J.H. Allender and T.P. Barnett, 1985: Computations and parameterizations of the nonlinear energy transfer in a gravity-wave spectrum - Part 2. J. Physical Oceanography, 15, 1378-1390.

Jacobs, S., 1987: An asymptotic theory for the turbulent flow over a progressive water wave. J. Fluid Mechanics, 174, 69-80.

Jeffreys, H., 1924: On the formation of water waves by wind. Proc. Royal Soc. A, London, 107, 189-206.

Jeffreys, H., 1925: On the formation of water waves by wind. (second paper). Proc. Royal Soc. A, London, 110, 341-347.

Longuet-Higgins, M.S., 1962: Resonant interactions between two trains of gravity waves. J. Fluid Mechanics, 12, 321-332.

Miles, J.W., 1957: On the generation of surface waves by shear flows. J. Fluid Mechanics, 3, 185-204.

Miles, J.W., 1959: On the generation of surface waves by shear flows - Part 2. J. Fluid Mechanics, 6, 558-582.

Miles, J.W., 1967: On the generation of surface waves by shear flows - Part 5. J. Fluid Mechanics, 30, 163-175.

Mistsuasu, H. and T. Honda, 1982: Wind-induced growth of water waves. J. Fluid Mechanics, 123, 425-442.

Munk, W.H., G.R. Miller, F.E. Snodgrass and N.F. Barber, 1963: Directional recording of swells from distant storms. Phil. Trans. Royal Soc., A, London, 255, 505-584.

Phillips, O.M., 1957: On the generation of waves by turbulent wind. J. Fluid Mechanics, 2, 417-445.

Phillips, O.M., 1959: The scattering of gravity waves by turbulence. J. Fluid Mechanics, 5, 177-192.

Phillips, O.M., 1960: On the dynamics on unsteady gravity waves of finite amplitudes. J. Fluid Mechanics, 9, 193-217.

Phillips, O.M., 1963: On the attenuation of long gravity waves by short breaking waves. J. Fluid Mechanics, 16, 321-332.

Phillips, O.M., 1985: Spectral and statistical properties of the equilibrium range in wind-generated gravity waves J. Fluid Mechanics, 156, 505-531.

Plant, W.J., 1982: The relationship between wind stress and wave slope. J. Geophysical Research, 87, C3, 1961-1967.

Riley, D.S., M.A. Donelan and W.H. Hui, 1982: An extended Miles' theory for wave generation by wind. Boundary-Layer Meteorology, 22, 209-225.

Roll, H.U., 1951: Neue Messungen Zur Enstehung von Wasserwellen durch wind. Ann. Meteor., 4, 269-286.

Shemdin, O.H. and E.Y. Hsu, 1967: The dynamics of wind in the vicinity of progressive water waves. J. Fluid Mechanics, 32, 497-531.

Snyder, R.L., 1974: A field study of wave-induced pressure fluctuations above surface gravity waves. J. Marine Research, 32, 497-531.

Snyder, R.L. and C.S. Cox, 1966: A field study of the wind generation of ocean waves. J. Marine Research, 24, 141-178.

Snyder, R.L., F.W. Dobson, J.A. Elliot and R.B. Long, 1981: Array measurements of atmospheric pressure fluctuations above surface gravity waves. J. Fluid Mechanics, 102, 1-59.

Stokes, G.G., 1880: Supplement to a paper on the theory of oscillatory waves. Mathematical and Physical papers, 1, Cambridge University Press, 314-326.

Toba, Y., 1973: Local balance in the air-sea boundary processes. III. On the spectrum of wind waves. J. Oceanogr. Soc. of Japan, 29, 209-220.

CHAPTER 4
WAVE PREDICTION: EARLY WAVE PREDICTION TECHNIQUES

4.1 General comments

Although a lot of brilliant mathematical analysis on wave theory was done in the mid and late nineteenth century, very little operational wave analysis work was reported during this time. This lack of interest (or progress) in operational wave analysis may have been due to two reasons: First, systematic ocean wave observations and recording was a difficult and an expensive operation and, second, ocean waves are too untidy to be mathematically attractive. As can be seen in Figure 4.1a, a sea-wave record produced by a standard wave recorder looks very irregular and chaotic compared to a sine wave record (Figure 4.1b) produced in a laboratory. Lord Rayleigh, after inspecting some of the wave records remarked, "The basic law of the seaway is the apparent lack of any law".

During the early days of navigation, mariners and sailors had developed a wind-scale with appropriate description of the sea-state at different steps of the scale. In 1805, British Rear-Admiral Sir Francis Beaufort developed a numbering system which was applied to the various steps of the mariners' descriptive wind scale; this numerical scale was adopted for general use in the British Navy in 1834. In 1903, a scale of equivalent wind speed was introduced by a formula,

$$U = 1.87B^{3/2} \qquad (4.1)$$

Here U is the wind speed in miles per hour and B is the Beaufort number. In recent years, the custom of judging wind force by the appearance of the sea surface has been accepted internationally. The commission for Marine Meteorology of the World Meteorological Organization (WMO, Geneva, Switzerland) has prepared a Marine Science Affairs Report (WMO, 1970) which provides a Table linking Beaufort numbers with corresponding sea state descriptions and the wind speed values. For ready reference, the Beaufort Scale of wind force as proposed by WMO is reproduced here in Table 4.I; this Beaufort Scale has been accepted internationally for reporting sea-state conditions.

The Beaufort scale with its associated wave-height values was the only operational procedure available for describing sea-state in

TABLE 4.I: BEAUFORT SCALE OF WIND FORCE

Beaufort Number	Descriptive term	Interval of equivalent wind speeds*		Mean equivalent wind speed*	Brief specifications for observing at sea	Probable wave height	Probable maximum wave height
		knots	m s^{-1}	m s^{-1}		m	m
0	Calm	0 - 2	0 - 1.3	0.8	Sea like a mirror	0	0
1	Light air	3 - 5	1.4 - 2.7	2.0	Ripples with the appearance of scales are formed, but without foam crests	0.08	0.0
2	Light breeze	6 - 8	2.8 - 4.5	3.6	Small wavelets, still short but more pronounced; crests do not break	0.15	0.1
3	Gentle breeze	9 - 12	4.6 - 6.6	5.6	Large wavelets; crests begin to break; foam of glassy appearance	0.6	0.9
4	Moderate breeze	13 - 16	6.7 - 8.9	7.9	Small waves, becoming longer; fairly frequent white horses	1.1	1.5
5	Fresh breeze	17 - 21	9.0 - 11.3	10.2	Moderate waves, taking a more pronounced long form; many white horses	1.8	2.6
6	Strong breeze	22 - 26	11.4 - 13.8	12.6	large waves begin to form; white foam crests more extensive everywhere	2.9	4.0
7	Near gale	27 - 31	13.9 - 16.4	15.1	Sea heaps up and white foam from breaking waves begins to be blown in streaks along the direction of the wind	4.1	5.8
8	Gale	32 - 37	16.5 - 19.2	17.8	Moderately high waves of greater length; edges of crests begin to break into the spindrift; foam is blown in well-marked streaks along the direction of the wind	5.5	7.6

TABLE 4.I : BEAUFORT SCALE OF WIND FORCE (Cont'd)

Beaufort Number	Descriptive term	Interval of equivalent wind speeds*		Mean equivalent wind speed*	Brief Specifications for observing at sea	Probable wave height	Probable maximum wave height
		knots	$m\ s^{-1}$	$m\ s^{-1}$		m	m
9	Strong gale	38 – 43	19.3 – 22.4	20.8	High waves; dense streaks of foam along the direction of the wind; crests begin to topple, tumble and roll over	7.0	9.8
10	Storm	44 – 50	22.5 – 26.0	24.2	Very high waves with long overhanging crests; resulting foam, blown in dense white streaks along the direction of the wind; visibility affected	8.8	12.5
11	Violent storm	51 – 57	26.1 – 30.0	28.0	Exceptionally high waves (small and medium-sized ships might be for a time lost to view behind the waves); sea is completely covered with long white patches of foam lying along the direction of the wind	11.3	15.8
12	Hurricane	≥ 64	≥ 32.6	32.2	The air is filled with foam and spray; sea completely white with driving spray; visibility very seriously affected	13.7	

* Applicable to observations made on board ship. The equivalent wind speeds correspond to an anemometer height of 20 meters above sea surface.

The last two columns are from Allan(1983); these columns are added as a guide to show roughly what may be expected in the open sea, away from land. The probable maximum wave height is reached in about one wave in ten.

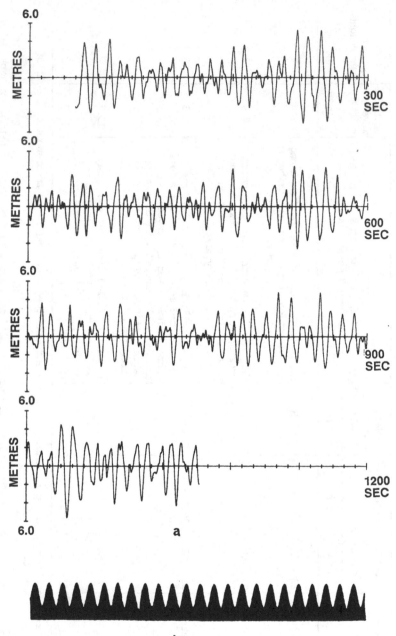

Figure 4.1: a. An example of a sea wave record from a wave recorder (from the archives of Marine Environmental Data Service, Ottawa). b. A sine wave record produced in a laboratory.

the early 1900s. The serious study on wave analysis and prediction
began during the World War II, in response to a crucial need for wave
forecasts for planning of amphibious operations. In 1942, the U.S.
Navy commissioned Drs. H. Sverdrup and W. Munk to develop forecasts of
sea-state conditions over the North Sea. This led to the first operational wave prediction procedure which was developed in 1943 and
reported later in 1947; the theoretical development and the operational procedure are briefly described below:

4.2 Significant Wave Method

The pioneering study of Sverdrup and Munk(1947) used an important physical innovation by considering both the tangential and the
normal stress components in the wave generation mechanism. It may be
recalled that the earlier sheltering hypothesis theory of Jeffreys
(1924) had neglected the tangential stress and considered the normal
stress as the only important stress for wave generation. According to
Sverdrup and Munk, the condition for wave growth is:

$$R_T \pm R_N > R_\mu \qquad (4.2)$$

Here R_T and R_N are the tangential and normal components of the wind
stress and R_μ is the rate of wave energy dissipation due to molecular
viscosity. Substituting appropriate expressions for R_T and R_N, we can
rewrite Equation (3.3) developed earlier, as

$$2\rho_a C_D U^2 c \pm s\rho_a (U - c)^2 c > 4\mu g$$

This equation allows the growth of waves even when the phase speed c
is more than the wind speed. When the wave phase speed exceeds the
wind speed, the relative wind velocity will be directed against the
direction of the wave motion and the sign of R_N will be -ve in eq.
(4.2). For wave phase speed less than the wind speed, eq. (4.2) will
be used with a +ve sign.

Next in order to apply the basic equations of wave dynamics,
Sverdrup and Munk used a statistical description of the sea state by
introducing <u>a parameter 'significant wave height' which is defined as
the mean of the highest one third of the waves present in the sea at
any given time</u>. The significant wave height represents the height
which an experienced observer will come up with while estimating a
characteristic wave height for a given sea-state condition. Sverdrup
and Munk further obtained the energy balance equation by considering
the energy budget of the significant and the conservative (or the
classical) waves. The energy balance equation is written as

$$\frac{dE}{dt} + \frac{E}{c}\frac{dc}{dt} = R_T \pm R_N \qquad (4.3)$$

Here E refers to the mean energy of the wave per unit surface area. In obtaining equation (4.3), the condition $\frac{\partial E}{\partial x} = 0$ is applied to the significant waves under the assumption that the significant waves do not vary in a horizontal direction when a uniform wind blows over an open ocean. Equation (4.3) may be defined as the <u>Duration equation</u> in a sense that the integration of this equation will give the change with time of the significant waves at any position in a storm area which generates the waves.

Initially the significant waves will have originated in the immediate vicinity of a point under observation. As time increases, the waves near the observation point would have come from farther and farther away. In a real situation, the distance from which waves can come at an observing location is limited by the dimensions of a storm (which produces the waves) or by the presence of a shoreline in an upwind direction. The time necessary for the waves to travel from the beginning of the fetch to the point of observation may be defined as the minimum duration (t_{min}). Beyond this t_{min}, there will be no growth with time of the waves at the point of observation; in this case, a steady-state can be said to have been reached ($\frac{\partial E}{\partial t} = 0$) and we can write an equation,

$$\frac{c}{2} \frac{dE}{dx} + \frac{E}{2} \frac{dc}{dx} = R_T \pm R_N \qquad (4.4)$$

Equation (4.4) is called the <u>Fetch equation</u> as the solution to this will provide wave speed and height as a function of the fetch after the steady-state has been reached, i.e. for $t \geq t_{min}$.

Equations (4.3) and (4.4) are the two basic equations of the Sverdrup-Munk technique. In order to solve these equations, they are rewritten in a nondimensional form using the parameters δ(wave steepness = H/L) and β(wave age = c/U). Assuming δ to be a function of β only, Equations (4.3) and (4.4) can be rewritten as

$$\text{the duration equation :} \quad \frac{d\beta}{dt} = f(\delta, \beta, U)$$

$$(4.5)$$

$$\text{the fetch equation} \quad : \quad \frac{d\beta}{dx} = f(\delta, \beta, U)$$

The functional form of the equations (4.5) can be seen in Sverdrup and Munk(1947). In order to develop a relationship between δ and β, Sverdrup and Munk gathered all available data from different sources and plotted the same to examine the relationship between δ and β. As can be seen from Figure 4.2, the wave steepness δ reaches a maximum value of about 0.1 when the wave age β is about 0.4; the wave steepness drops off somewhat for younger waves ($\beta < 0.4$) and diminishes

rapidly for older waves. The solid line drawn through the data points
is the assumed relationship between the significant wave steepness and
wave age. From this, other relationships such as wave height and speed
as a function of fetch and duration are obtained. For practical purposes, equations (4.5) are solved in terms of the following dimensionless parameters:

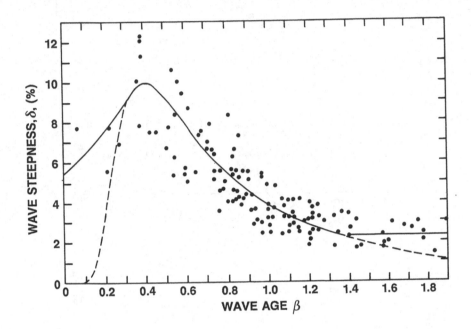

Figure 4.2: Relationship between δ(wave steepness) and β (wave age)
as proposed by Sverdrup and Munk(1947)

$$\beta(= \frac{c}{U}) \quad \text{for wave speed} : \quad \frac{gH}{U^2} \quad \text{for wave height}$$

$$\frac{gx}{U^2} \quad \text{for the fetch} : \quad \frac{gt}{U} \quad \text{for the duration}$$

(4.6)

Here U is the wind speed at the anemometer level and g is the gravitational acceleration. Further, the solutions are obtained for three
different ranges of β as follows:

$$0 \leq \beta \leq \beta' = 0.350 ; \quad \beta' \leq \beta \leq 1 ; \quad 1 \leq \beta$$

Having obtained the solutions, graphs of β against the nondimensional
fetch parameter $\frac{gx}{U^2}$ and against the nondimensional duration parameter
$\frac{gt}{U}$ were prepared by Sverdrup and Munk. These graphs which are reproduced in Figure 4.3(a,b) are called the Fetch graph and the Duration

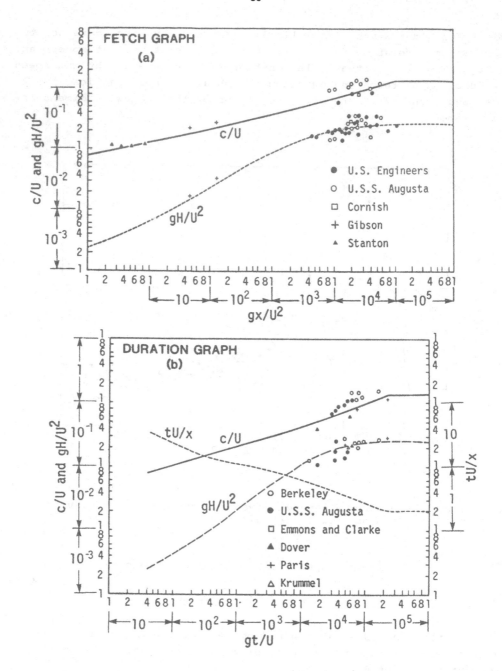

Figure 4.3: (a) Fetch graph and (b) Duration graph as obtained by Sverdrup and Munk.

graph respectively. The curves in the graphs show the theoretical relationships, while the observations from various sources are shown by symbols. From the fetch graph (Figure 4.3a) one can obtain the wave phase speed over a given fetch when the wind has been blowing at a constant speed for an unlimited length of time, while Figure 4.3b gives wave height and phase speed for a given duration if the wind has been blowing at a constant speed over an unlimited fetch. In reality, both duration and fetch will be limited and the wind will not have a constant speed; as such, the values of wave height and phase speed obtained from Figure 4.3a will not in general agree with those obtained from Figure 4.3b and the smaller of the two is used. If the wind has blown at a constant velocity for a long time over a small body of water, the growth of waves at a location is found to depend entirely on the distance of that location from the upwind shore. On the other hand, if the wind has blown for a short time over a very long fetch, the wave growth will be primarily governed by the duration for which the wind has blown. Thus the sea is either fetch-limited or duration-limited depending on whether the fetch or the duration imposes greater restriction on the growth of the waves; physically, this means that the sea-state development is limited by the geometry of the ocean basin and the amount of total energy which the sea surface can extract from the wind under given atmospheric flow patterns.

The analysis of Sverdrup and Munk was extensively modified by Bretschneider in a series of papers (1951, 1952a, 1959, 1970, 1973). Bretschneider analyzed a large amount of field and laboratory data and developed semi-empirical wave forecasting relationships whose graphical solutions can be seen in one of the well-known Bretschneider nomograms as displayed in Figure 4.4; this nomogram is based on the following wave forecasting relationships:

$$\frac{gH_s}{U^2} = A_1 \tanh \left[B_1 \left(\frac{gF}{U^2} \right)^{m_1} \right]$$

$$\frac{c}{U} = \frac{gT_s}{2\pi U} = A_2 \tanh \left[B_2 \left(\frac{gF}{U^2} \right)^{m_2} \right]$$

(4.7)

In the above, H_s and T_s are the significant wave height and wave period, F is the fetch length, U is the average surface wind at 10 m height above water level, c is the phase speed of the significant wave and A_1, A_2, B_1, B_2, m_1 and m_2 are parameters obtained empirically by Bretschneider(1973). The nomogram gives significant wave height and period as a function of wind speed, fetch length and duration assuming that the wind speed etc. do not change with time.

The use of Bretschneider nomogram can be illustrated by the

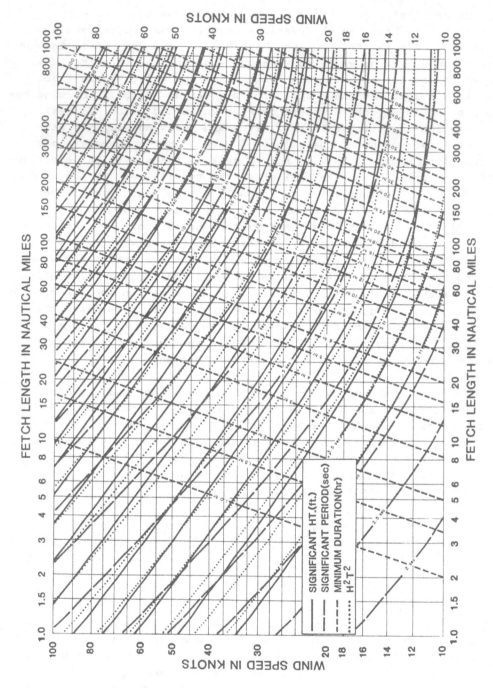

Figure 4.4: Bretschneider's (1970) nomogram for wave analysis and forecasting (reproduced with permission from Prof. Bretschneider)

following example: Given that wind speed U = 30 knots, wind duration
t = 12 hr and over-the-water fetch F = 200 nautical miles (nm) at a
location; obtain the sea-state conditions at that location.

To solve this problem, enter the Bretschneider nomogram
(Figure 4.4) at U = 30 knots and move in the x-direction until the
given value of F(fetch) or t(duration) is reached. In this example,
the duration curve t = 12 is reached first at which point

$$\text{significant wave height } H_s = 11 \text{ ft}$$
$$\text{significant wave period } T_s = 7.4 \text{ s}$$

At this point in the nomogram, the fetch value can be read as
F = 110 nm; this is identified as the minimum fetch length (f_{min})
required for a 30 knot wind to generate a wave of 11 feet in 12 hours.
Next, assume that the wind continues to blow for an additional 12
hours, i.e. total duration t = 24 hr. In this case, F = 200 nm is
reached first on the horizontal line U = 30 kt; at that intersection,
we have

$$\text{significant wave height } H_s = 14 \text{ ft}$$
$$\text{significant wave period } T_s = 8.3 \text{ s}$$

Further, from the duration lines, one obtains t_{min} = 20 hr. Here the
waves are fetch-limited because after t = 20 hr, there is no increase
in the wave height for the same fetch length.

Once the waves have travelled out of the storm area where they
are generated, they continue to travel as swells over long distances.
These swells lose energy through lateral diffraction, spreading due to
wave dispersion, air-resistance and wave-current interaction. Sverdrup
and Munk developed wave decay relationships based on a limited amount
of data. Bretschneider(1952b) revised these relationships using additional data on generation and decay of wind waves and obtained suitable nomograms showing wave height and period as a function of the
decay distance. Figure 4.5 shows these nomograms giving variations in
wave period and wave height with increasing decay distance. From these
nomograms, the decay in wave height and wave period can be read off
for a given value of the decay distance.

The above example and discussion illustrate the well-known
SMB(Sverdrup, Munk, Bretschneider) technique in which atmospheric data
(wind speed, wind duration etc.) are used to diagnose and predict
sea-state conditions at a given oceanic location. Several other
nomogram-based wave forecasting techniques were developed and reported
in the 1940's and 1950's; noteworthy among them are the nomograms by

54

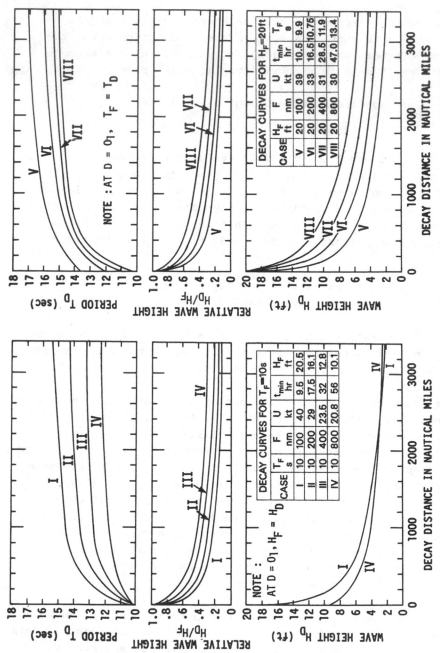

Figure 4.5: Wave decay curves. The ordinate gives the significant wave period (T_D) and the significant wave height (H_D at the end of the decay distance (D). T_F and H_F are respectively, the significant wave period and significant wave height at the end of fetch F. The curves on the left show relationship between $T_F = 10$ s, F = 100, 200, 400 and 800 nm, decay and wave period at end of decay. The curves on the right show relationship between $H_F = 20$ ft, F = 100, 200, 400 and 800 nm, decay and wave height at end of decay (from Bretschneider, 1952, AGU Transactions, vol. 33, copyright by the American Geophysical Union).

Suthons(1945) of the British Admiralty, by Darbyshire and Draper
(1958) of the Institute of Oceanography (England) and by Groen and
Dorrestein of the Netherlands Meteorlogical Institute (see WMO, 1976).
These nomogram-based techniques can provide operationally useful
results in situations where wind speed and wind duration could be
assumed constant. In reality, wave producing weather systems often
create variable wind speed, durations and fetch lengths. In such a
situation, a moving fetch method developed by Wilson(1955,1963) is
found to be better suited. Wilson used equations (4.7) with constants
A_1, A_2 -- derived empirically and solved them by a stepwise numerical
integration scheme to obtain solutions of H_S and T_S for an assumed
steady wind speed U acting over a small fetch Δx. Solving these equations over a wide range of wind speeds and wind duration enabled
Wilson to develop a complex wave forecasting nomogram. The Wilson's
nomogram provides an improved solution of wave height and wave period
in a moving wind system with variable wind speed. A computerized procedure based on Wilson's moving fetch method has been developed and
put into an operational wave prediction system by the AES, Canada;
more details of this wave prediction system are given in Chapter
Nine.

4.3   Wave Spectrum Method

The techniques discussed in the previous section obtain wave
characteristics of a 'significant wave' which is defined in a statistical sense. The significant wave concept does not take into account
the spectral character of the sea-state as revealed by the irregular
oscillations of the wave record of Figure 4.1a. This spectral character of the sea-state was being increasingly realized which led to
the development of the wave spectrum method of Pierson, Neumann and
James(1955). The wave spectrum method predicts the spectrum of a wave
from which the significant wave height and other wave parameters can
be derived. According to this PNJ (Pierson, Neumann, James) technique,
a function that describes mathematically the distribution of the
square of the wave height with frequency is called the spectrum of the
wave motion; since the square of the wave height is related to the
potential energy of the sea surface, the wave spectrum is also called
the energy spectrum.

For a given wind speed, the sea-state can be imagined to consist of an infinite number of wave components having a range of frequencies. Let the infinite number of wave components be approximated
by a finite number of sine waves with slightly differing average frequencies f_i, centered around a frequency interval Δf; further, let
the square of the amplitude of each of these frequencies be denoted
by $A^2(f_i)$. The wave spectrum can be schematically represented by a

'stairway' approximation as shown in Figure 4.6a; here the ordinate is the square of the amplitude of the wave with an associated frequency f_i. The area of each rectangle in Figure 4.6a is proportional to the square of the wave height and this in turn is proportional to the wave energy per unit sea surface area contained in each wave train of average frequency f_i. The continuous unimodal curve shows the limiting form of the wave spectrum for a given wind speed. The smooth unimodal spectrum of Figure 4.6a may be considered as an idealized representation of the sea state at a given location.

In Figure 4.6b are shown three continuous wave spectra for a fully developed sea at wind speeds of 20, 30 and 40 knots respectively. A fully developed sea at a given wind speed is defined when all possible components in the spectrum between the frequencies f = o and f = ∞ are present with their maximum amount of spectral energy. It can be seen from Figure 4.6b that the maximum frequency band displaces itself towards lower frequencies as the wind speed increases. An empirical formula connecting wind speed (U) and maximum frequency (f_m) for a fully developed sea is given by

$$f_m U = 2.467 \qquad (4.8)$$

In (4.8), f_m is given in Hz and U in knots.

An important fact of the wave spectrum is that as the spectral wave energy in a broad band around the peak frequency (f_m) increases with increasing wind speed, the contribution of wave energy at the tail ends of the curves (in Figure 4.6a) becomes less and less significant, compared with the amount of wave energy concentrated around the peak frequency. Thus, wave components with very high or very low frequencies can be neglected as they do not noticeably affect the dominating wave pattern of the sea; accordingly, for operational wave forecasting, the wave spectrum can be conveniently cut off at high and low frequencies for values below a certain amount of spectral wave energy. This is the basis of the PNJ technique which obtains cocumulative wave spectra over only the significant part of the wave spectrum. The cocumulative spectrum can be constructed from the wave spectrum as shown in Figure 4.7. The ordinate of the cocumulative spectrum gives the value of $E(m^2$ or $ft^2)$ which is defined as twice the variance of a large number of values from points equally spaced in time as chosen from a wave record. The top part of Figure 4.7 represents a wave spectrum for a fully developed sea, while the hatched area of the spectrum represents the total wave energy of individual wave components up to and including the wave frequency f. The numerical value of the hatched area is given by the ordinate E_a of the cocumulative graph shown in the lower half of Figure 4.7. At f_c, almost all the area under the

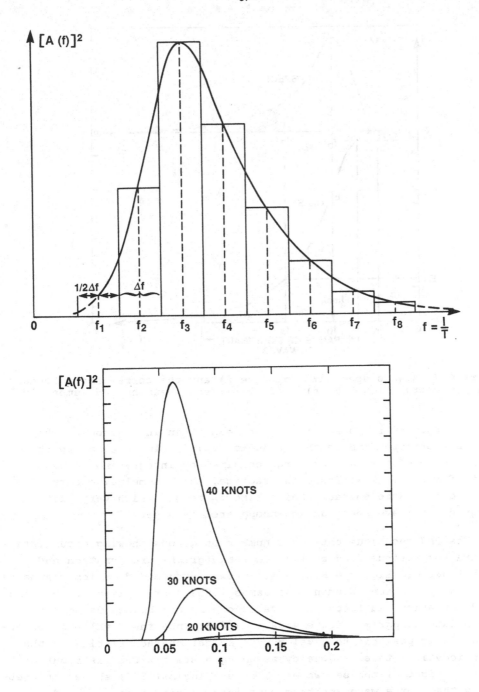

Figure 4.6: a. A 'stairway' approximation to a wave spectrum b. wave
spectra for a fully developed sea at wind speeds of 20, 30 and 40
knots (from Pierson, Neumann and James, 1955).

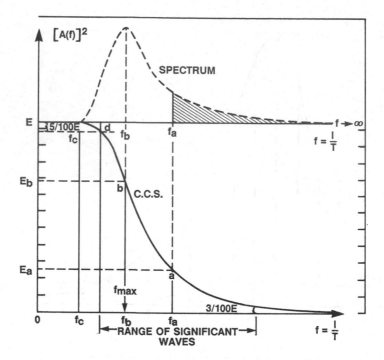

Figure 4.7: A wave spectrum (upper half) and the corresponding cocumulative spectrum (lower half). (from Pierson, Neumann and James, 1955)

spectrum between f = o and f = ∞ has been taken into account. The value E_c(corresponding to the frequency f_c) is almost equal to the total energy of the wave spectrum obtained by integrating between limits f = o and f= ∞. Thus, in practice, the cocumulative curve can be cut off at some suitable point like f_c beyond which very little energy of the wave spectrum is encountered in general.

The PNJ technique obtains a number of graphs showing cocumulative spectra for various wind speeds. On these graphs are duration and fetch lines. In Figure 4.8(a,b) are shown fetch and duration graphs as obtained by Pierson, Neumann and James(1955). For a given set of wind speed, duration and fetch the E value can be determined using the appropriate nomogram (Figure 4.8a or 4.8b). Once the E value is determined, it is possible to develop wave height characteristics at the given location; this is done by assuming a statistical distribution for the height of the sea waves. Longuet-Higgins(1952) showed theoretically that for a wave spectrum containing a single narrow band of frequencies, the wave height H follows the Rayleigh distribution for which the probability density function p(H) is given by,

$$p(H) \ dH \ = \ \frac{H}{4m} \ \exp(- \ \frac{H^2}{8m}) \ dH \tag{4.9}$$

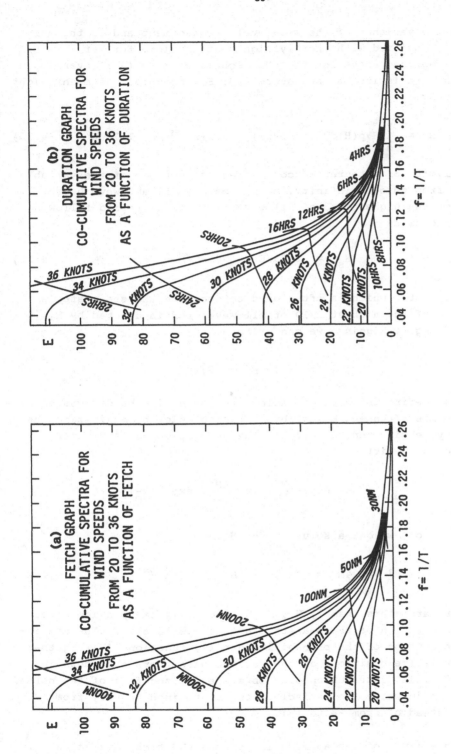

Figure 4.8: a. Co-cumulative spectra for wind speed as a function of fetch (Fetch graph). b. Co-cumulative spectra for wind speed as a function of duration (Duration graph). [from Pierson, Neumann and James, 1955]

Here m is the variance of the sea-level displacement and is the only parameter associated with the Rayleigh distribution; further, m = E/2 where E is defined above for the PNJ technique. From (4.9), various wave height statistics can be determined. For example, the mean height $\bar{H}$ is given by

$$\bar{H} = \int_0^\infty H \; p(H)dH = (2\pi m)^{\frac{1}{2}} = (\pi E)^{\frac{1}{2}} = 1.77\sqrt{E} \qquad (4.10)$$

Here the limits of integration cover the entire range of the variable H, namely from zero to infinity. Next, the significant wave height H_s (mean of the highest 1/3 of all waves) can be expressed following Longuet-Higgins as

$$H_s = 2.83\sqrt{E} \simeq 4.00\sqrt{m} \qquad (4.11)$$

Another statistic which is often used to describe the sea-state is the average of the highest 1/10 of all waves and is denoted by $H_{1/10}$. Following Longuet-Higgins, we can write,

$$H_{1/10} = 3.60\sqrt{E} \simeq 5.10\sqrt{m} \qquad (4.12)$$

We can also define the most probable wave height (to be designated as H_m) as the most frequent H and obtain its solution by maximizing the probability density function (4.9). The derivative of (4.9) with respect to H is written as,

$$\frac{dp}{dH} = \frac{1}{4m} \exp\left(-\frac{H^2}{8m}\right) - \frac{2H^2}{32m^2} \exp\left(-\frac{H^2}{8m}\right)$$

For $\dfrac{dp}{dH} = o$, we get a solution for H as

$$H_m^2 = 4m = 2E \; ; \quad \rightarrow \quad H_m = 1.41\sqrt{E} \qquad (4.13)$$

These four parameters, namely H_m, $\bar{H}$, H_s and $H_{1/10}$ can provide a reasonable description of the sea-state at a given location. The statistical distribution of the probability density function p(H) together with the relative positions of the various parameters are shown in Figure 4.9. This distribution of H is a specialized form of the normal distribution in which the variable H is constrained to vary from 0 to ∞ (instead of from $-\infty$ to $+\infty$).

As an example of the application of the PNJ technique, consider the earlier problem in which U = 30 kt, wind duration t = 12 hours and

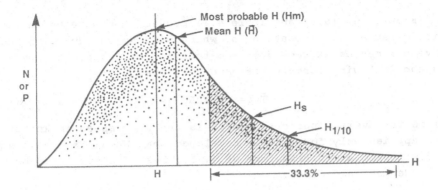

Figure 4.9: The statistical distribution of wave heights showing various parameters (from Bretschneider, 1964)

fetch F = 200 nm. Using the Figure 4.7(a,b), we obtain E = 38 ft² from the fetch graph and E = 15 ft² from the duration graph; this suggests that the sea is not fully developed and accordingly we use the smaller of the two E values namely E = 15 ft². With this value, we can use formulas (4.10) to (4.13) to obtain;

most frequent wave height	= 5.5 ft
average wave height	= 6.9 ft
significant wave height	= 11.0 ft
average of highest 1/10 of the waves	= 19.8 ft.

In this example, the waves are said to be duration limited. In situations when the sea is either duration limited or fetch limited, the sea is not considered to be fully developed. There is a set of minimum fetch and duration values needed to generate a fully developed sea for a given wind speed; these values of minimum fetch and duration for selected wind speeds are given in Table 4.II.

TABLE 4.II: Minimum fetch and duration values needed to generate a fully developed sea for various wind speeds (from Pierson, Neumann and James, 1955)

U	knot	10	20	30	40	50	56
F_m	nm	10	75	280	710	1420	2100
t_m	hr	24	10	23	42	69	88

Figures 4.7(a,b) and Table 4.II provide the wave height characteristics at a given location for which wind speed, duration and fetch values can be extracted from meteorological data. For wave period the following formula has been developed:

$$\bar{\bar{T}} = 0.285U \qquad (4.14)$$

Here $\bar{T}$ is the average period(s) and U is the wind speed in knots; this formula applies only for a fully developed sea. When the sea is not fully developed, a modified form of (4.14) has been suggested by Pierson, Neumann and James.

The wave spectrum method of PNJ as described above may be interpreted as a progressive step in the direction of spectral wave model development. The PNJ technique may have provided an impetus that led to a number of theoretical and experimental studies relating wave spectrum formulations. Some of the important wave spectrums that have been proposed in the last thirty years are briefly described in the following section.

4.4 Wave Spectrum Formulation

a. Neumann (1953): Using a large number of individual wave observations, Neumann computed the parameters $\frac{H}{T^2}$ and $\frac{T}{U}$ and arrived at the following relation,

$$\frac{H}{gT^2} = \text{const. } \exp(-\frac{gT}{2\pi U})^2 \qquad (4.15)$$

Here the parameter $\frac{H}{T^2}$ is directly related to the wave steepness and $\frac{T}{U}$ to the wave age. The above relation can be expressed in terms of energy density as a function of wave frequency f as,

$$E(f) = \text{const. } f^{-6} \exp[-2f^2g^2U^{-2}] \qquad (4.16)$$

Here E(f) is the energy of the wave component corresponding to the frequency f.

b. Phillips(1958): Through a dimensional analysis, Phillips derived an empirical relationship which predicts that at high frequencies the energy density decreases with increasing frequency as the inverse fifth power of the frequency; the functional form of the energy density is given by,

$$E(f) \sim \alpha g^2 f^{-5} \qquad \text{(inverse fifth power law)} \qquad (4.17)$$

The range of frequencies for which this relation holds is called the equilibrium range. Phillips' inverse fifth power law is meant to hold for frequencies larger than that of the peak frequency in the energy spectrum and smaller than those for which surface tension is important; there is a sizable frequency band which satisfies this condition. As mentioned earlier (section 3.4) Phillips(1985) has recently re-examined this relationship based on the present knowledge on wave-wave interaction, energy input from the wind and wave breaking. Assuming that all the three processes are important in the equilibrium range, Phillips proposes the frequency spectrum to be proportional to U_*gf^{-4} where U_* is the friction velocity; this revised expression by Phillips emphasizes the importance of U* in wind wave modelling.

c. Bretschneider(1959): Based on a large number of observations, Bretschneider proposed the following form for a design wave spectrum:

$$E(f) = Af^{-5} \exp[-Bf^{-4}] \tag{4.18}$$

Here A and B are defined in terms of significant wave height H_S and the modal (or the peak) period T_m; this spectrum or a variation of this defined by other similar equations is widely used in naval design work and in seakeeping operations (Bales et al. 1982; Comstock et al. 1980).

d. Pierson and Moskowitz(1964): Based on 420 selected wave measurements recorded with a ship-borne wave recorder and using a similarity theory of Kitaigorodskii(1961), Pierson and Moskowitz proposed the following form for a fully developed sea:

$$E(f) = \alpha g^2(2\pi)^{-4} f^{-5} \exp[-\beta(\frac{g}{Uf})^4] \tag{4.19}$$

Here U is the mean wind measured at 19.5 m level above the sea surface, $\alpha = 8.1 \times 10^{-3}$ and $\beta = 0.74$. This spectrum known commonly as the P-M spectrum has been found to be in close agreement with observed spectrum for wind speeds from 10 to 40 knots. Spectral wave models developed by Pierson et al(1966) use the P-M spectrum as the limiting spectrum in the energy balance equation.

e. JONSWAP (Joint North Sea Wave Project, 1973)

Observations made during the JONSWAP suggested that the wind-sea spectrum in the growing phase has a much sharper peak than the P-M spectrum. Accordingly, a JONSWAP spectrum has been defined by Hasselmann et al(1973) in the following form:

$$E(f) = \alpha g^2(2\pi)^{-4} f^{-5} \exp\{-\frac{5}{4}(\frac{f_m}{f})^4 + \ell n \ \gamma . \exp[-\frac{(f - f_m)^2}{2\sigma^2 f_m^2}]\} \tag{4.20}$$

Here σ has two values $\sigma \bigg\{ = \begin{array}{ll} \sigma_a & \text{for } f \leq f_m \\ \sigma_b & \text{for } f \geq f_m \end{array}$

The JONSWAP spectrum as given by (4.20) has five parameters; f_m the peak frequency and α the equilibrium range constant (or Phillips' constant) are called the scale parameters, while γ the peak enhancement factor, σ_a the left peak width and σ_b the right peak width are called the shape parameters. The JONSWAP spectrum is schematically shown in Figure 4.10. The parametric representation of the wind-sea spectrum is justified on the hypothesis that nonlinear wave-wave interactions control the evolution of the spectrum in a growing sea; further, the nonlinear interaction is believed to impart a stabilizing influence on the shape of the spectrum such that an invariant form can be assumed. Using this argument, Hasselmann et al(1976) have proposed a parametric wave prediction model which will be discussed in the next Chapter.

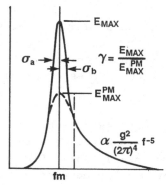

Figure 4.10: A schematic of the JONSWAP spectrum with its five parameters. E_{MAX} denotes the maximum energy for the JONSWAP spectrum while E_{MAX}^{PM} denotes the maximum energy for the corresponding Pierson-Moskowitz (P-M) spectrum. (from Hasselmann et. al. 1973)

f. The Wallops spectrum (1981): Based on a theoretical analysis and laboratory data collected at the NASA (National Aeronautics and Space Administration) Wallops Flight Centre in Virginia, U.S.A., Huang et al(1981) proposed a unified two parameter spectrum given by,

$$E(f) \quad = \quad \frac{\beta g^2}{f_m^5} \left(\frac{f}{f_m}\right)^{-n} \exp\left[-\frac{n}{4}\left(\frac{f}{f_m}\right)^{-4}\right] \tag{4.21}$$

The Wallops spectrum depends upon two parameters, β and n which are functions of the internal parameter namely the significant slope of the wave field. A comparison of the Wallops spectrum with the JONSWAP spectrum and the field data shows an excellent agreement in general. Huang et al further suggest the possibility of using remotely sensed

data to obtain an estimate of the internal parameter, the significant
slope. This study (by Huang et al.) has been extended by Liu(1983) who
has proposed a generalized form for a wind wave spectrum in terms of
three coefficients which can be determined from readily available
spectral parameters like energy E and the peak frequency f_m. Liu
further shows that the coefficients of the proposed spectrum are
closely related to the stages of the wave growth.

Both the above studies suggest interesting practical applications. The choice of Wallops spectrum offers the possibility of using
remotely sensed data as a direct input into the spectral model; the
generalized spectrum of Liu may be useful in developing an operational
wave prediction model. Besides these spectral forms, a few other forms
have been proposed in the last fifteen years or so (ex. Toba, 1973;
Kruseman,1976). The spectrum proposed by Kruseman is being used in an
operational wave prediction model at present. (see Chapter Five for
more details).

In summary, the development of various spectral forms over the
last thirty years or more has provided a sound basis for describing
the sea-state using a prescribed analytical form; this has led to the
development of modern spectral wave models, some of which are described in the following Chapter.

REFERENCES

Allen, W.T.R., 1983: Wind and sea: State of sea photographs for the
Beaufort wind scale. Env. Canada, Canadian Govt. Pub. Centre, Ottawa,
1983, 57 pp.

Bretschneider, C.L., 1951: The generation and decay of wind waves in
deep water. Tech. Rep. No. 155-46, Inst. of Engineering Research,
Univ. of California, Berkeley, Aug. 1951.

Bretschneider, C.L., 1952a: Revised wave forecasting relationships.
Proc. Second Conf. Coastal Engineering, Berkeley, U.S.A., 1-5.

Bretschneider, C.L., 1952b: The generation and decay of wind waves in
deep water. Trans. American Geophy. Union, 33, 381-389.

Bretschneider, C.L., 1959: Wave variability and wave spectra for wind-
generated gravity waves. Tech. Memo. No. 118, Beach Erosion Board,
U.S. Army Corps of Engineers, Aug. 1959, 192 pp.

Bretschneider, C.L., 1964: Wave generation by wind, deep and shallow
water. Estuary and Coastal Hydrodynamics, Engineering Societies Monograph. Ippan Arthur T. (ed.), McGraw Hill, U.S.A., Ch. 3, 133-196.

Bretschneider, C.L., 1970: Forecasting relations for wave generation.
Look Lab/Hawaii, 1, No. 3, University of Hawaii, U.S.A., Aug. 1970.

Bretschneider, C.L., 1973: Prediction of waves and currents. Look Lab/Hawaii, 3, No. 1, University of Hawaii, U.S.A., Jan. 1973.

Darbyshire, M. and L. Draper, 1963: Forecasting wind generated sea waves. Engineering, 195, 482-484.

Hasselmann, K., et al., 1973: Loc. cit. (Chapter 3).

Hasselmann, K., B. Ross, P. Müller and W. Sell, 1976: A parametric wave prediction model. J. Physical Oceanography, 6, 200-228.

Huang, N.E., S.R. Long, C. Tung, Y. Yuen and L. Bliven, 1981: A unified two-parameter wave spectral model for a general sea state. J. Fluid Mechanics, 112, 203-224.

Jeffreys, H., 1924: Loc. cit. (Chapter 3).

Kruseman, P., 1976: Two practical methods of forecasting wave components with period between 10 and 25 seconds near Hoek van Holland. Wetensch. Rapp. 76-1, Koninklijk Nederlands Meteorological Instituut, De Bilt, Netherlands.

Liu, P.C., 1983: A representation for the frequency spectrum of wind-generated waves. Ocean Engineering, 10, 429-441.

Longuet-Higgins, M.S., 1952: On the statistical distribution of the heights of the sea waves. J. Marine Research, 11, 245-266.

Neumann, G., 1953: An ocean wave spectra and a new method of forecasting wind generated sea. Tech. Memo. 43, Beach Erosion Board, U.S. Army Corps of Engineers.

Phillips, O.M., 1958: The equilibrium range in the spectrum of wind generated waves. J. Fluid Mechanics, 426-434.

Phillips, O.M., 1985: Loc. cit. (Chapter 3).

Pierson, W.J. and L. Moskowitz, 1964: A proposed spectral form for fully developed seas based on the similarity theory of S.A. Kitaigorodskii. J. Geophysical Research, 5181-5191.

Pierson, W.J., G. Neumann and R. James, 1955: Practical methods for observing and forecasting ocean waves by means of wave spectra and statistics. H.O. Pub. 603, U.S. Navy Hydrographic Office, Washington, D.C., 284 pp.

Pierson, W.J., L.J. Tick and L. Baer, 1966: Computer based procedures for preparing global wave forecasts and wind field analysis capable of using wave data obtained by a spacecraft. Proc. Sixth Naval Hydrodynamic Symposium, Washington, D.C., U.S.A., 499-533.

Suthons, C.T., 1945: The forecasting of sea and swell waves. Memo. 134/45, Naval Meteorological Branch, U.K.

Sverdrup, H.U. and W.H. Munk, 1947: Loc. cit. (Chapter 1).

Toba, Y., 1973: Loc. cit. (Chapter 3)

Wilson, B.W., 1955: Graphical approach to the forecasting of waves in moving fetches. Beach Erosion Board, U.S. Army Corps of Engineers, Tech. Memo. No. 73, 31 pp.

Wilson, B.W., 1963: Deep water wave generation by moving wind systems. Trans. Amer. Soc. Civil Eng., 128, paper No. 3416.

WMO, 1970: The Beaufort scale of wind force. World Meteorological Organization, Geneva, Marine Science Affairs, Report No. 3, 22 pp.

WMO, 1976: Handbook of wave analysis and forecasting. World Meteorological Organization, Geneva, No. 446.

CHAPTER 5
WAVE PREDICTION: SPECTRAL MODELS

5.1 General Comments

The framework of contemporary ocean wave prediction models is the spectral energy balance equation which can be written as,

$$\frac{\partial}{\partial t} E(f,\theta,\bar{x},t) \; + \; \bar{c}_g \cdot \nabla E \; = \; S(f,\theta,\bar{x},t) \tag{5.1}$$

Here E is the energy density of the wave field described as a function of frequency f, direction θ, position $\bar{x}$ and time t; c_g is the group velocity of the wave field in deep water and S is the source function representing the physical processes that transfer energy to and from the wave spectrum. The source function S can be symbolically written as,

$$S \; = \; S_{in} + S_{n\ell} + S_{ds} \tag{5.2}$$

Here S_{in} is the energy input to the wave field from the atmosphere, $S_{n\ell}$ represents the transfer of energy associated with nonlinear wave-wave interactions and S_{ds} represents the energy dissipation which includes dissipation in deep as well as in shallow waters. If the various physical processes represented by the terms in (5.2) can be accurately specified, equation (5.1) can be numerically integrated to yield wave prediction with an accuracy limited only by errors in wind specification and in numerical methods.

During the last thirty years, several important contributions have been made towards development of suitable expression for the various terms in (5.2). In particular, the term S_{in} has been formulated based on the wave generation theories of Phillips(1957) and Miles(1957). Most operational spectral wave models at present express the wind input as a sum of linear growth term of Phillips and an exponential growth term of Miles; these linear and exponential terms can be formulated using equations (3.4) and (3.5) respectively. The complete form of the nonlinear term $S_{n\ell}$ has been given by Hasselmann (1962) and others; recent spectral wave models have attempted inclusion of complete (or almost complete) form of $S_{n\ell}$ as we shall discuss later. The dissipation term S_{ds} has been the most difficult one to specify accurately, mostly due to lack of reliable data on wave

dissipation. In deep waters, most operational wave models have used
an exponential type term to represent wave dissipation. In shallow
waters, wave dissipation has been represented primarily by the bottom
friction effects and the wave breaking mechanism.

The various spectral wave models developed and reported so far
can be classifed into three broad catagories depending on the way the
model is formulated. In the following sections, some of these spectral
wave models are briefly summarized.

5.2 Discrete Spectral Wave Models

Models in this catagory involve representation of the directional spectrum by a discrete number of finite bandwidth spectral
components travelling in a specified number of directions; this provides a detailed sea-state information at a given location in terms of
a two-dimensional (frequency-direction) spectrum. The governing equation for these spectral models is given by (5.1) which can be solved
by standard finite difference techniques over a series of time-steps
on a grid covering the ocean basin of interest.

One of the earliest discrete spectral models based on eq.
(5.1) was reported by Gelci et al(1957) and was called 'Densité
spectro-angulaire (DSA) model'. The DSA model was based on three
source terms: a theoretical advection term, a generation term and a
dissipation term, the last two being empirically fitted to reproduce
observed fetch and duration dependence. The first version of the DSA
model provided predictions at a single point in which space-time wind
fields were obtained from manually plotted ship wind observations and
a propagation diagram was used for calculation of advection and rate
of change of spectral energy. The spectral growth was 'simultaneous'
and all components were assumed to grow independently, to their
asymptic equilibrium value given by the 'Pierson-Neumann' spectrum.
Later in 1960, with the introduction of computers, a numerical grid
was designed for the DSA model and an improved damping term was used;
subsequent modifications of the DSA model and related results have
been reported by Gelci et al(1964) and Gelci and Chevy(1978). It must
be remembered that at the time when the DSA model was first developed,
the wave generation theories of Phillips and Miles were not reported
in literature and very little information was available on the other
two terms, namely $S_{n\ell}$ and S_{ds}.

The pioneering efforts of Prof. W.J. Pierson (New York University, U.S.A.) has led to a class of spectral wave models which use the
Pierson-Moskowitz (P-M) spectrum as the limiting spectrum. An operational spectral wave model developed by Pierson, Tick and Baer(1966)

known commonly as the PTB model uses the growth equation in the following form:

$$\frac{\partial}{\partial t} E_w(f_i) = A(f_i, U) + B(f_i, U_*) \cdot E_w(f_i) \tag{5.3}$$

The right side of (5.3) is a functional representation of S_{in}, the energy input from the atmosphere. In this representation, A is based on Phillips' linear growth mechanism, B is based on Miles-Phillips' exponential growth mechanism, U is the (surface) wind speed and U_* is the friction velocity as defined by eq. (3.8). The symbol $E_w(f_i)$ represents the energy in the ith frequency band summed over all directions within $\pm 90°$ to the wind direction, θ_w; that is,

$$E_w(f_i) = \sum_j E(f_i, \theta_j) \quad \text{where } \theta_w - 90 < \theta_j < \theta_w + 90 \tag{5.4}$$

Here $E(f_i, \theta_j)$ is the energy value of the wave field at frequency f_i and direction θ_j.

The growth equation (5.3) is modified using the concept of a fully developed sea; that is, it is assumed that if the wind blows uniformly in speed and direction over a sufficiently large area and for a sufficiently long period of time, the wave spectrum in that area will attain the fully developed form of the P-M spectrum as given by eq.(4.19). Based on this limiting state, nonlinear effects that would act during wave generation are modeled implicitly by modifying (5.3) according to

$$\frac{\partial}{\partial t} E = [A\{1 - (\frac{E}{E_\infty})^2\}^{\frac{1}{2}} + BE][1 - (\frac{E}{E_\infty})^2] \tag{5.5}$$

Here E refers to $E_w(f_i)$ as expressed in (5.3), E_∞ is the value of the P-M spectrum for frequency f_i for a given wind speed U and the factor $[1 - (\frac{E}{E_\infty})^2]$ plays an important part in modelling the transition from a growing sea to a fully developed sea. The expression for E_∞ is further modified by inclusion of a directional spreading factor $G(f, \theta^*, U)$ so that,

$$E_\infty = \alpha g^2 (2\pi)^{-4} f^{-5} \exp\{-\beta(\frac{g}{Uf})^4\} [G(f, \theta^*, U)] \tag{5.6}$$

In (5.6), the spreading factor G has a functional form involving $\cos 2\theta^*$ and $\cos 4\theta^*$, where θ^* is measured with reference to the wind direction; the spreading factor is assigned a zero value whenever $|\theta^*| > \frac{\pi}{2}$. Other functional forms for the spreading factor have been suggested in literature. For example, Mitsuasu et al (1975) have suggested a spreading factor having terms in $(\frac{\cos \theta^*}{2})^p$. For values of p around 16, the spreading factor becomes narrow and has a tall sharp peak; such a distribution for the spreading factor is appropriate for low frequencies of the wave spectrum.

Equation (5.5) is the growth equation of the PTB model. The growth portion of the program operates at a given time step by the summation of contribution of the spectrum for a given f_i travelling within 90° to the wind. If $E_w(f_i) > E_\infty(f_i)$, there will be no change in the values of $E(f_i, \theta_j)$, but if $E_w(f_i) < E_\infty(f_i)$, the growth is computed at each frequency by solving (5.5) to obtain the incremental growth which is then spread out in different directions using the spreading factor G.

The Operational Spectral Ocean Wave Model (SOWM) of the U.S. Navy is a revised version of the PTB model (see Pierson, 1982). The SOWM operates in three steps: Grow, Dissipate and Propagate. In the first step, the solution of (5.5) with initial conditions E = o at t = o is expressed as,

$$E(f,t) = A\left[\frac{e^{Bt} - 1}{1}\right]\left[B^2 + \left\{\frac{A(e^{Bt} - 1)}{E_\infty}\right\}^2\right]^{-\frac{1}{2}} \qquad (5.7)$$

At the start of the time step, $t = t_o$, the spectral band has reached the value $E_w = E_w(f_i) = \sum_{\theta_w} E(f_i, \theta_j)$ which represents the sum over directions of all elements in the spectral array at a given frequency and within 90° of the wind direction θ_w. If E_w exceeds E_∞, the spectral band remains unchanged. However, if $0 < E_w < 0.95 E_\infty$, that frequency component of the wind-generated sea must grow to a new value during the next time step. The time t_o required for the spectral component to have grown to the value $E_w[\text{or } E_w(f_i)]$ needs to be found out first; this can be done using eq. (5.7) where the left hand side is $E(f_i, t_o) = E_w$. Solving for t_o, we get,

$$e^{Bt_o} = 1 + \frac{BE_w}{A(1 - E_w^2/E_\infty^2)^{\frac{1}{2}}} = R(A, B, E_w, E_\infty) \qquad (5.8)$$

It can be seen from (5.8) that t_o can be different for different spectral bands. The new value of the spectral component at $t_o + \Delta t(\Delta t$ is the time step) is given by (5.7) in which t is replaced by $t_o + \Delta t$. Substituting from (5.8) and simplifying, we can get an expression for the increase in the spectrum during Δt as,

$$\Delta E(f_i) = E(f_i)_{New} - E_w(f_i)$$

$$= \frac{A[Re^{B\Delta t} - 1]}{B\left[1+\left(\dfrac{A(Re^{B\Delta t} - 1)}{BE_\infty}\right)^2\right]^{\frac{1}{2}}} - E_w \qquad (5.9)$$

Here R is given by (5.8). Equation (5.9) gives the increase in the spectral component as a function of time step Δt, A, B, E_w, and E_∞. Further, if $0.95E_\infty \le E_w(f_i) < E_\infty$, then

$$\Delta E(f_i) = E_\infty(f_i,U) - E_w(f_i) \qquad (5.10)$$

The increment in the frequency spectrum is obtained either from eq. (5.9) or from (5.10); this increment is then spread out in various directions according to the spreading factor G referred to in equation (5.6).

Next, the growth terms A and B are formulated based on the theoretical studies of Phillips and Miles. The linear growth term A is calculated as a function of the frequency f_i and wind speed at 6.1 m level above the water surface; the wind at 6.1 m level is obtained from the input-level (19.5 m) wind by assuming neutral stability in the atmospheric boundary layer. The exponential growth term B is expressed as a function of the frequency f_i and the friction velocity U_* (see Pierson, 1982 for more details).

The second step of the SOWM is the dissipation step in which waves travelling against the wind are attenuated according to the following expression,

$$E_d(f_i, \theta_j) = E_0(f_i,\theta_j)[exp(-Cf_i^4\sqrt{E_w})]^{K(\theta_j)} \qquad (5.11)$$

Here E_d is the spectral component after dissipation, E_0 is the spectral component before dissipation and $K(\theta_j)$ is the dimensionless power whose value depends upon the direction of the wave component travelling relative to the wind direction. (for wave components travelling nearly opposite to the wind, the value of $K(\theta_j)$ is 6 while for wave components travelling nearly at right angles to the wind, it is equal to 3). The value of the constant C was chosen as 690 and was determined by analyzing wave records obtained by a British weather ship during a sequence of extratropical storms. In deriving (5.11), wave breaking is considered to be the most important process in attenuating waves. The turbulence associated with breaking waves is assumed to have its greatest impact on high frequency waves which are rapidly eliminated with the use of fourth power for the frequency in the exponential coefficient of (5.11).

The third step of the SOWM is the propagation step in which each spectral component is propagated along great circle arcs at a deep-water group velocity appropriate to its frequency. Most numerical modelling studies in atmosphere and ocean dynamics use comformal map projections on which a suitable grid is chosen to obtain numerical integration of governing equations. Grid paths on such projections are not in general along great circle arcs, although for distances less than a quarter of the earth's surface, deviations from the great circle paths are small and can be neglected. The SOWM uses a grid based on an icosahedral-gnomonic projection (see Figure 5.1) so that all lines joining arrays of grid points are along great circle arcs. Each of the twenty triangular subprojections has 325 grid points and for the northern hemisphere over which the SOWM operates, there are about 1575 ocean points. The propagation of spectral energy on such a grid is accomplished along six primary and six secondary directions. Further, to simulate the sudden arrival of a swell, a discontinuity field is created and a jump technique developed by Baer(1962) is utilized; more details of the propagation scheme are given by Pierson (1982).

The two-dimensional spectrum used in the SOWM consists of 12 direction bands (30° resolution) and 15 frequency bands with central frequencies for individual bands ranging from 0.308 to 0.038 (corresponding to periods from 3.2s to 25.7s). The SOWM has been used operationally by the U.S. Navy since December 1974 to produce wave forecasts for the three ocean basins of the northern hemisphere. Since June 1985, the SOWM has been replaced by the GSOWM (Global Spectral Ocean Wave Model) which is a global model having the same number of frequency bands as SOWM but twice the number of direction bands. The GSOWM operates on a standard 2.5° latitude by 2.5° longitude spherical grid over a global band extending from 77.5°N to 72.5°S. The GSOWM uses essentially the same phyics for wave growth, dissipation and angular spreading as its predecessor, SOWM. The GSOWM is the world's first global operational wave forecast model (Clancy, Kaitala and Zambresky, 1986). Another model based on the PTB model formulation is the ODGP (Ocean Data Gathering Program) model, developed by Cardone et al.(1976). The ODGP model uses a finer grid spacing than the SOWM in deep water and half of that spacing on the continental shelf region. In addition, the ODGP uses 24 direction bands and 15 frequency bands and has an improved version of the wave growth algorithm. The ODGP model was calibrated against hurricane wind and wave data in the Gulf of Mexico; however, the ODGP model was found to work quite well for extratropical storms as well. The nested grid of the ODGP model is shown in Figure 5.2; this grid covers the western north Atlantic region over which the model is used in an operational mode at present by Dr. V.J. Cardone (of Oceanweather Inc., U.S.A.). During the 1986

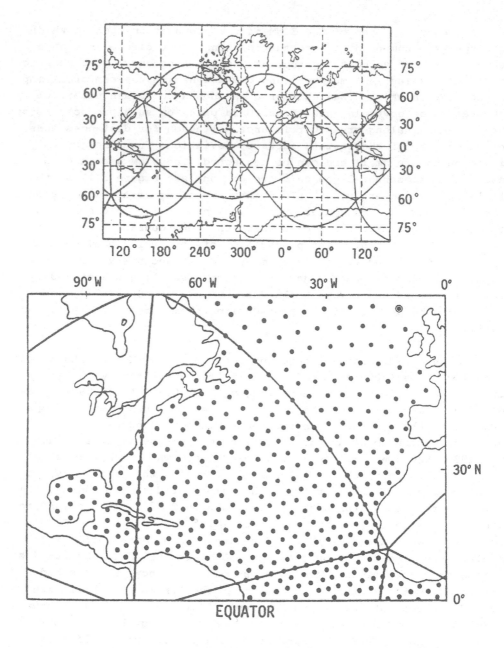

Figure 5.1: a. The twenty equilateral triangles of the Icosahedral Gnomonic projection for the Spectral Ocean Wave Model (SOWM). b. Arrangement of gridpoints for a portion of the SOWM grid covering the North Atlantic ocean. The circled grid point is the location where some of the SOWM products were evaluated, as discussed in Chapter 7 (from Pierson, 1982).

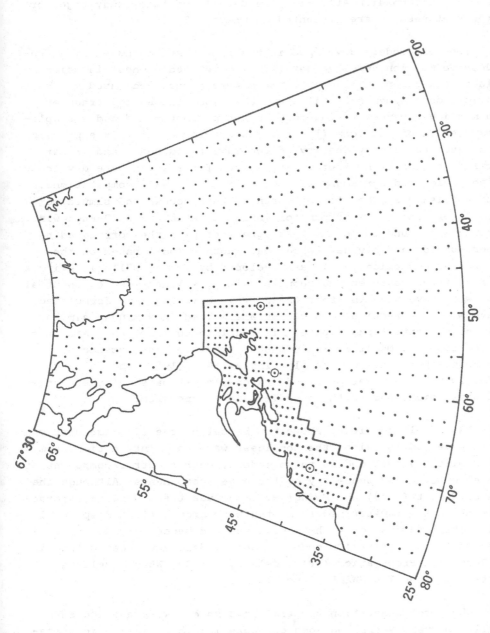

Figure 5.2: The Operational ODGP model grid. The coarse grid has a spacing of 1.25° latitude by 2.50° longitude. The nested fine grid has half the spacing of the coarse grid. Three circled grid points are the locations for wave model inter-comparison study. (source: Dr. V. Cardone)

field project of the Canadian Atlantic Storms Program (CASP), the ODGP model was included in a wave model intercomparison study which was designed to evaluate wind and wave products from several operational models over the Canadian Atlantic. The details of this study together with relevant results are presented in Chapter 7.

Two more models developed in the late sixties and early seventies deserve mention here. Barnett(1968) developed a model in which the linear and exponential growth terms were formulated based on the theoretical developments of Miles and Phillips. The energy transfer due to nonlinear wave-wave interaction was parameterized and a simplified expression of the form (Γ - rE) was developed; in this expression E is the spectral energy density of the wave field and Γ and r are integral functions of E. Appropriate values for Γ and r were developed so as to yield good agreement with Hasselmann's (1963) computations. Barnett applied his model to the north Atlantic ocean and simulated the growth of wind-sea spectrum for infinite fetch and a 30-knot wind; the growth was found to be in close agreement with the corresponding P-M spectrum for a fully developed sea. Another model based on the energy balance equation (5.1) was developed by Ewing(1971). In Ewing's model four terms representing respectively, the linear and exponential wave growth, wave-wave interaction and wave breaking were formulated; the wave-wave interaction term was parameterized in a manner similar to that of Barnett. Ewing applied his model to the northeast Atlantic Ocean and compared model wave hindcasts with wave measurements at two ocean weather stations, namely, India (59°N, 19°W) and Juliett (52.5°N, 20°W); the model provided reasonable estimates of the significant wave height and of the one-dimensional spectrum.

The models mentioned above may be called the first generation spectral wave models. Since the nonlinear wave-wave interaction term is not explicitly included in these models, each spectral component of the model evolves independently of the other components. Although the models of Barnett(1968) and Ewing(1971) include the nonlinear interaction terms in a parameterized form, the wave growth and the spectral form are still dominated by the wind input and hence they are classified as the first generation models. These models have been catagorized as Decoupled Propagation (DP) models by the Sea WAve Modelling Project-SWAMP (see The SWAMP Group, 1985).

The first generation spectral wave models were applied successfully for many years and some of these models are still in operation today. (ex. GSOWM of the U.S. Navy). However, these models were found to underestimate the observed wave growth requiring augmentation of the A and B terms in equation (5.3). These models were further unable to explain the overshoot phenomenon of a growing wind sea first

observed by Barnett and Wilkerson(1967) and subsequently confirmed by
other workers. According to Barnett and Wilkerson, the growth of a
particular spectral component when traced along the fetch in a generating area showed a rapid increase, overshooting the equilibrium value
and then dropping sharply, undershooting to a minimum value. This
overshot and undershoot effect is shown schematically in Figure 5.3.

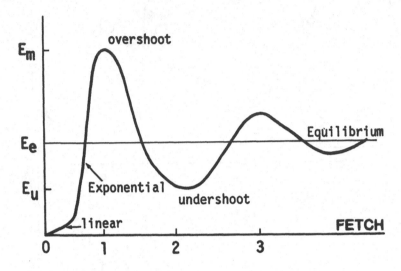

Figure 5.3: A schematic growth curve of a spectral component with
fetch showing overshoot and undershoot effect. The fetch is shown in
nondimensional units, while E_m, E_e and E_u denote the maximum, equilibrium and underequilibrium energy values (from Mitsuasu,1982).

        The overshoot effect is now considered to be a real phenomenon
that forms an integral part of the wave generation process and is
believed to be caused by the nonlinear energy transfer. During the
JONSWAP field experiment (Hasselmann et al. 1973), the evolution of
wave spectrum with fetch was measured by several wave riders located
along a wave array in the North Sea (see Figure 5.4a,b). For a selected case of offshore winds, the evolution of the one-dimensional spectrum at various wave rider locations is shown in Figure 5.5; these
wave spectra clearly show the rapid growth of wave energy on the forward (low frequency) face of the spectrum and this growth is primarily
associated with the nonlinear energy flux across the peak due to
resonant wave-wave interactions. Further, it was found that the nonlinear energy transfer controlled not only the rate of growth of the
newly developing waves, but also the form of the spectrum, in particular the development of the pronounced peak and the migration of the
peak toward lower frequencies. These important findings have led to
the development of second and third generation spectral wave models
which are summarized in the following three sections:

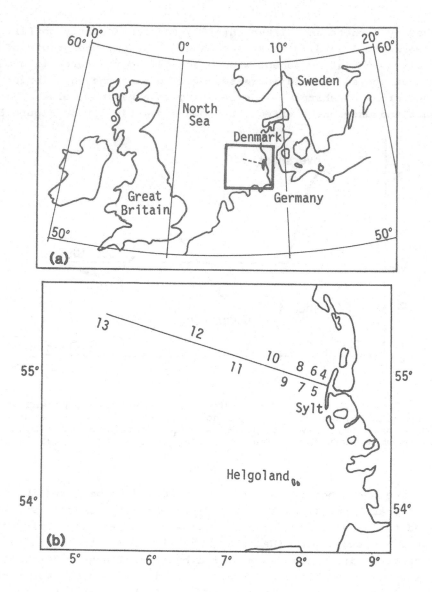

Figure 5.4:   a. The site of the JONSWAP field experiment (1968, 1969).
b. The JONSWAP wave array extending 160 km into the North Sea, west-
ward from Sylt, Denmark. The numbers refer to the wave measuring locations (from Hasselmann et al. 1973).

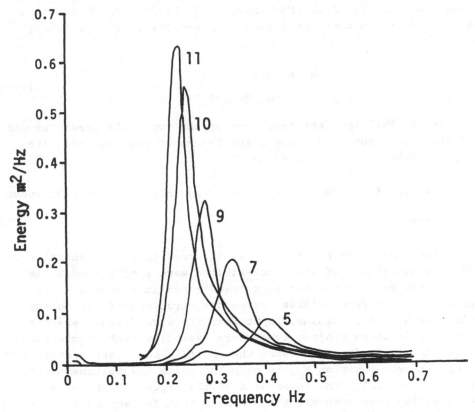

Figure 5.5: Evolution of the observed wave energy spectrum with fetch for offshore winds (15 Sept. 1968, 1100-1200h). The numbers refer to the wave measuring locations along the JONSWAP wave array. (from Hasselmann et al. 1973)

5.3 Parametric and Hybrid Wave Models

According to Hasselmann et al(1976), nearly all fetch-limited frequency spectra measured during the JONSWAP field experiment could be fitted closely by the JONSWAP spectrum given by (4.20). This spectrum is derived from the P-M spectrum (equation 4.19) by multiplying with a peak enhancement factor,

$$\gamma \exp[-\frac{(f - f_m)^2}{2\sigma^2 f_m^2}] \tag{5.12}$$

This enhancement factor is reduced to zero for $\gamma = 1$. Since γ is the ratio of the peak value of the JONSWAP spectrum to the peak value of the P-M spectrum, the value 1 for γ makes the two spectra identical. The JONSWAP data as well as other field and laboratory data from many sources (ex. Burling, 1959; Mitsuasu, 1968; Pierson and Moskowitz, 1964) were found to obey the following power-law relations fairly closely;

$$\nu = 3.5\ \xi^{-0.33}$$
$$\alpha = 0.076\ \xi^{-0.22} \tag{5.13}$$

Here α is the Phillips' constant (one of the two scale parameters of the JONSWAP spectrum) and ν and ξ are the nondimensional peak frequency and fetch values defined as:

$$\nu = Uf_m/g\ ; \qquad \xi = gx/U^2 \quad (f_m: \text{peak frequency; } U: \text{mean wind speed}$$

at 10 m level) $\tag{5.14}$

From the JONSWAP data, it was inferred that the general form of the energy balance of the fetch-limited wave spectrum must lie somewhere between the two limiting cases shown in Figure 5.6. The upper half of the Figure (5.6a) shows the structure of the energy balance for a case of minimal input into the wave field; here the dissipation is taken minimum everywhere except at high frequencies where a sink is needed to balance the positive nonlinear transfer. The nonlinear source function $S_{n\ell}$ is computed using the Boltzmann integrals (ex. eq. 3.10). The lower half of the Figure (5.6b) shows the other limiting case when all the momentum transfer across the air-sea interface is wave induced. In this case, the increased input in the central region of the spectrum for moderate and large fetches must be balanced by a dissipation term S_{ds} which is calculated using a whitecapping model of Hasselmann(1974). In both cases, the shape of the spectrum is independent of the detailed distribution of the input and is primarily controlled by the nonlinear energy transfer for various spectral shapes. Further, the spectrum adjusts to a self-stabilizing form which is continually maintained by the nonlinear transfer.

Based on the JONSWAP data as well as other field data, the mean values of the three shape parameters of the JONSWAP spectrum were found to be;

$$\gamma = 3.3\ ; \qquad \sigma_a = 0.07\ ; \qquad \sigma_b = 0.09 \tag{5.15}$$

Using the above as specified values of the three shape parameters, Hasselmann et al(1976) proposed a parametric wave model in which the form of the spectrum is governed by equation (4.20), the JONSWAP spectrum. Of the five parameters required to define the JONSWAP spectrum, the three shape parameters were prescribed by eq.(5.15); for the re-

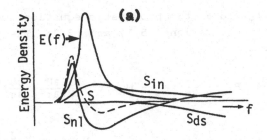

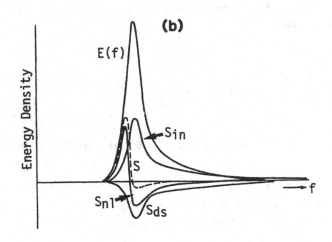

Figure 5.6: The structure of spectral energy balance for _a_. a minimum input into the wave field and negligible dissipation in the main part of the spectrum (from Hasselmann et al. 1973); _b_. a maximum input into the wave field (from Hasselmann, 1974; Copyright by D. Reidel Publishing Company)

maining two parameters f_m and α, two prognostic equations were derived based on the energy balance equations (5.1) and (5.3). The two prognostic equations were:

$$\frac{1}{\nu} \left(\frac{\partial \nu}{\partial \tau} + P_{\nu\nu} \frac{\partial \nu}{\partial \eta}\right) + P_{\nu\alpha} \frac{1}{\alpha} \frac{\partial \alpha}{\partial \eta} = -N_\nu \alpha^2 \nu + \frac{1}{U}\left(\frac{\partial U}{\partial \tau} + \frac{\partial U}{\partial \eta}\right)$$

$$\text{(5.16)}$$

$$\frac{1}{\alpha} \left(\frac{\partial \alpha}{\partial \tau} + P_{\alpha\alpha} \frac{\partial \alpha}{\partial \eta}\right) + P_{\alpha\nu} \frac{1}{\nu} \frac{\partial \nu}{\partial \eta} = I\nu^{7/3} - N_\alpha \alpha^2 \nu + \frac{0.2}{U}\left(\frac{\partial U}{\partial \eta}\right)$$

In the above equations, ν is the nondimensional peak frequency defined earlier in (5.14). The various coefficients in equation (5.16) have following values;

$$\begin{bmatrix} P_{\nu\nu} & P_{\nu\alpha} \\ P_{\alpha\alpha} & P_{\alpha\nu} \end{bmatrix} = \begin{bmatrix} 1 & -0.07 \\ 0.47 & 0.2 \end{bmatrix} \; ; \; N_\nu = 0.54, \; N_\alpha = 5,$$

$$I = 5.1 \times 10^{-3}$$

Further, the partial derivatives in (5.16) are defined as:

$$\frac{\partial}{\partial \tau} = \left(\frac{U}{g}\right) \frac{\partial}{\partial t}; \; \frac{\partial}{\partial \eta} = \left(\frac{U}{g}\right) \bar{V}_m \cdot \nabla \; ; \; \bar{V}_m \text{ is parallel to the wind direc-}$$

tion and is given by $|\bar{V}_m| = gq/4\pi f_m$ where $q = 0.85$. The dimensionless gradient $\frac{\partial}{\partial \eta}$ corresponds to the rate of advection of properties with the group velocity $\bar{V}_m$ of waves in the spectral peak; the correction factor q arises from averaging over the directional distribution of the spectrum.

In (5.16), the two terms on the right side of the first equation represent the contribution from the nonlinear interaction ($S_{n\ell}$) and the nonuniform wind field source functions respectively; the three terms on the right side of the second equation in (5.16) represent the contribution from the wind input (S_{in}), the nonlinear interaction ($S_{n\ell}$) and the nonuniform wind field source functions respectively. In general, equations (5.16) could be solved using finite difference analogues of these equations on a spatial grid on which time and space varying wind fields could be prescribed. Hasselmann et al(1976) proposed a simplification in which a quasi-equilibrium first order solution for α was expressed as;

$$\alpha = B\nu^{2/3} \; ; \; B = \left(\frac{I}{N_\alpha}\right)^{\frac{1}{2}} = 0.032 \qquad \text{(5.17)}$$

The above relationship between α and ν was hypothesized on the basis of a quasi-equilibrium between the nonlinear transfer and the combined

influence of atmospheric input and dissipation. With this (5.17)
relationship, the wave prediction problem was reduced to solving only
the first equation of (5.16) for the parameter ν. The JONSWAP data as
well as other available field data appeared to support the relationship between α and ν as expressed by (5.17), while the prognostic
equation for ν was found to give reasonable results only under uniform
wind and limited fetch conditions. Thus the parametric wave model as
described by equation (5.16) and (5.17) is not applicable when the
wind field is rapidly changing or when the sea-state is transitional.
Further, the model is limited to growing wind seas and hence cannot be
applied in situations when the sea is swell-dominated.

The parametric wave model represents a significant conceptual
advance in wave modelling techniques because it allows the nonlinear
wave-wave interactions to be included without involving the time
consuming calculations of the Boltzmann integrals describing the nonlinear interactions. However, the parametric model as described by
Hasselmann et al has limited applicability. In view of this, wave
models which combine the parametric wave representation with discrete
representation have been developed and reported by Günther et al(1979)
and Janssen et al(1984) among others. These models have been classified as Coupled Hybrid (CH) models by The SWAMP Group(1985). The model
developed by Günther et al(1979) called NORSWAM (NORth Sea WAve Model)
is a typical example of a coupled hybrid ocean wave model which is
described below:

The NORSWAM uses the JONSWAP spectrum with two parameters
(σ_a and σ_b) prescribed using the JONSWAP data. For the remaining three
parameters namely f_m, α and γ the energy source terms are given in
the following form:

1.   Source term for f_m: This is dominated by nonlinear wave-
wave interaction and hence prescribed on theoretical ground,

$$\text{Source } (f_m) \;=\; -0.586\; \alpha^2 f_m^{\,2}\; \frac{(\gamma - 1)}{2\cdot 3} \qquad \text{for } \gamma > 1$$

$$\hspace{4cm} = \; 0 \hspace{3cm} \text{for } \gamma < 1 \tag{5.18}$$

2.   Source term for α: involves contribution from wave-wave
interaction and other sources, principally the atmospheric input;

$$\text{Source } (\alpha) \;=\; -5.0\; \alpha^3 f_m \;+\; k\nu^p \alpha f_m \tag{5.19}$$

Here $\nu = f_m U/g$ is the nondimensional peak frequency while k and p are
the principal free parameters of the model and are calibrated by com-

parison with observed wave conditions. The term $k\nu^p\alpha f_m$ is so chosen because a quasi-equilibrium relation is implied between α and ν (see equation 5.13).

3. <u>Source term for γ</u>: dominated by wave-wave interactions;

$$\text{source } (\gamma) \quad = \quad -16.0 \; \alpha^2 f_m (\gamma - \gamma_o)$$

here $\quad \gamma_o \quad = \quad 3.2787 \qquad\qquad$ for $\; \nu \geq 0.16$

$\qquad \gamma_o \quad = \quad -14.95 + 113.94\nu \qquad$ for $0.16 > \nu > 0.14 \qquad$ (5.20)

$\qquad \gamma_o \quad = \quad 1 \qquad\qquad\qquad\qquad$ for $\nu \leq 0.14$

This source term for γ ensures transition to a fully developed state ($\gamma \to 1$) as $\nu \to 0.14$ ($\nu = 0.14$ gives a value of f_m corresponding to the P-M spectrum). Having formed the source terms, three prediction equations similar to (5.16) can be formulated which can be solved by numerical techniques to yield predicted values of these parameters. This parameterical approach can be applied only to the wind-sea region of the energy spectrum. Under swell conditions, the atmospheric input is too low to support the spectrum at a level at which wave-wave interactions are effective. A shape stabilizing dynamic balance does not exist and the swell has to be treated as a freely propagating wave field. However, there are situations when the swell exists at frequencies such that energy is received from the atmosphere although the swell is outside the range of wave-wave interactions and is not absorbed into any existing wind-sea. In order to model this situation, it is assumed that the swell at a frequency f will not be absorbed into a wind-sea with a peak frequency f_m if $f_m < 0.9f_m$. On the other hand, energy will be received from the atmosphere if the swell frequency $f > f_o = \frac{g}{2\pi U\cos\theta}$ where θ is the wave direction. Thus, for $f_o < f < f_m$, the swell energy is allowed to grow with an atmospheric input function of the form;

$$S(f,\theta)\begin{cases} = \quad \dfrac{\pi f}{10} \; (\dfrac{f}{f_o} - 1) \; \dfrac{\rho_a}{\rho_w} \cdot e(f,\theta) \qquad \text{for } f > f_o \\[2mm] = \quad 0 \qquad\qquad\qquad\qquad\qquad\qquad \text{for } f < f_o \end{cases} \qquad (5.21)$$

Here ρ_a and ρ_w are densities of air and water respectively and $e(f,\theta)$ is the angular distribution of the wave spectrum. Further, in combining the wind-sea and swell regions, two simplifying assumptions are made: <u>1</u>, overall energy is conserved and <u>2</u>, the wave-wave interaction is such that the coupling and the decoupling of wind-sea and swell is rapid when it occurs. With these assumptions, the wind-sea $\leftrightarrow$ swell transition is carried out in the model using an empirical limiting value for the peak frequency. (see Weare and Worthington, 1979 for additional details).

The set of coupled partial differential equations of the model is solved by expressing the equations in finite difference form on a grid covering the North Sea and the eastern Atlantic Ocean (see Figure 5.7a) with an average grid spacing of about 100 km. A second order Lax-Wendroff method is employed to integrate the equations. The swell field is represented on a set of ray paths which in the absence of refraction follow great circles; these ray tracks are very nearly straight lines on the stereographic projection which is used for the model. Figure 5.7b shows the characteristic ray paths used in the model to represent the swell field. Only four ray paths are shown in Fig. 5.7b, the other four being exactly opposite to those shown in the Figure. A transformation between the characteristic ray grid and the cartesian swell grid becomes necessary when the transfer from swell-sea to wind-sea occurs. For the NORSWAM model, the discrete spectral components are calculated using ten frequency bins between 0.05 and 0.15 Hz and eight direction bins each at 45° interval.

The NORSWAM was used to obtain wave hindcasts for a set of 42 storms over the North Sea during the ten year period from 1966 to 1976. The model generated wave heights were compared with wave records at positions FAMITA and STEVENSON (see Figure 5.7a). By adjusting the α source term (equation 5.19), the bias in the significant wave height H_S predicted by the model was minimized; this gave a relationship between α and ν of the form;

$$\alpha = 0.032 \ \nu^{0.657} \tag{5.22}$$

The above relation is in close agreement with equation (5.17) which was obtained for JONSWAP data. Further, an excellent correlation is obtained between the observed and the predicted significant wave heights. Additional results on the evaluation of the model are presented by Ewing et al.(1979).

As an extension of the NORSWAM, a HYbrid PArametrical (HYPA) model has been recently proposed by Günther and Rosenthal(1985); the HYPA model includes an additional parameter θ_0 defined as the mean wind-sea direction. An analysis of the NORSWAM results indicates a systematic deviation between observed and model results for small-scale veering wind fields sampled during the JONSWAP field experiment; this deviation appears to have been caused by the crude directional assumption included in the NORSWAM. To rectify this systematic deviation, Günther et al(1981) defined a directional distribution function such that;

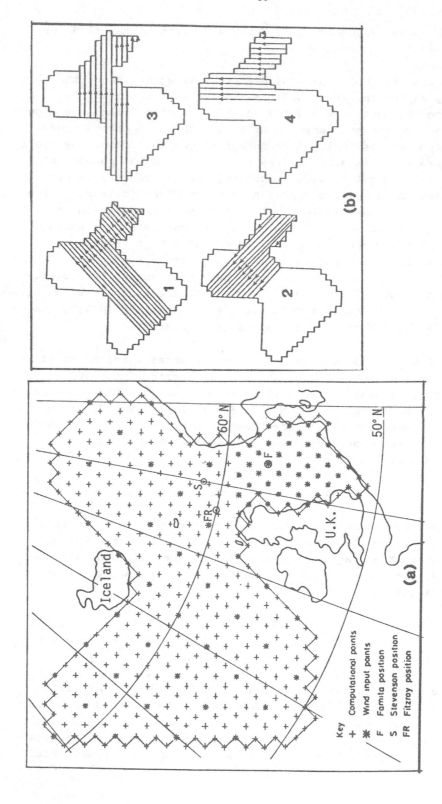

Figure 5.7: a. The grid used for the coupled hybrid model NORSWAM. b. The characteristic ray paths used for the swell field of NORSWAM. (from HRS report, 1977)

$$R(f,\theta) \begin{cases} = \frac{2}{\pi}\cos^2(\theta - \theta_o) & \text{for } |\theta - \theta_o| \leq \frac{\pi}{2} \\ \\ = 0 & \text{for } |\theta - \theta_o| \geq \frac{\pi}{2} \end{cases} \tag{5.23}$$

Here θ is the direction of propagation and θ_o is the mean direction of the wind-sea. The mean direction θ_o is obtained initially by calculating the direction of the total wave momentum vector and the source term for θ_o is obtained using a JONSWAP-type spectrum and assuming the peak frequency f_m to be defined as the lowest frequency in the experimental data that responds to the change in the wind direction; this source term for θ_o enables the wind-sea to turn into the new wind direction. The relaxation time for turning the mean wind direction has been estimated by Günther et al(1981) to be about six hours for a wind speed of 20 m s^{-1} and a peak frequency of 0.1 Hz. The swell propagation and calculations for the HYPA model are done exactly the same way as those for NORSWAM. The HYPA model has been recently upgraded to include shallow-water effects; this upgraded model called HYPAS (HYbrid PArametrical Shallow) has been tested at selected wave-rider locations in the North Sea. Some of the results of this testing will be presented in Chapter 7.

A spectral wave model developed at the Royal Netherlands Meteorological Institute (KNMI), DeBilt (The Netherlands), has been reported by Janssen et al(1984); this model called GONO (GOlven NOordzee) is another example of a Coupled Hybrid model which uses a finite difference scheme for the propagation of the parametrical wind-sea and a spectral ray technique for the propagation of swell. The spectral distribution of the wind-sea energy is given by

$$E_{wind\ sea} = \frac{2}{\pi}\cos^2(\theta - \phi)\ E(f)\quad |\theta - \phi| < \frac{\pi}{2}$$

Here ϕ is the wind direction, θ is any orbitrary direction and $E(f)$ is the Kruseman spectrum which is expressed as follows:

$$E(f) \begin{cases} = 0 & \text{for } 0 \leq f \leq f_{min} \\ \\ = \frac{\hat{\alpha}\ g^2}{(2\pi)^4} \cdot \frac{1}{f_m^5} \cdot \frac{f - f_{min}}{f_m - f_{min}} & \text{for } f_{min} < f < f_m \\ \\ = \frac{\hat{\alpha}\ g^2}{(2\pi)^4} \cdot \frac{1}{f^5} & \text{for } f > f_m \end{cases} \tag{5.24}$$

Here $\hat{\alpha}$ is Phillips' 'constant', f_{min} is the minimum frequency and f_m is the peak frequency of the spectrum. The Kruseman spectrum has been chosen over the JONSWAP spectrum because of its simpler analytical form. The parameters f_m and f_{min} are calculated using the total energy expression for the Kruseman spectrum and a 'stage-of-development' parameter.

The GONO model uses the basic energy balance equation (5.1) to calculate the advective part of the wave energy using a first-order finite-difference scheme. After each advection step, the average direction of propagation of the new energy in each grid point is calculated. This new energy is reduced proportionately depending on the difference between the average propagation direction and the local wind direction; this reduced energy is taken as the new wind-sea in the local wind direction. The remaining energy is considered as potential swell and is recovered by the swell energy scanning technique; this scanning technique allows the swell and the wind-sea energy to be combined and provides a reasonable simulation for a distant localized source in addition to a local wind sea. The swell calculations are done in six directional sectors and seven period bands with centre periods ranging from 7 to 19 s. The swell and wind-sea energy calculations are done in such a way so as to define a mean direction θ_0 for the wind-sea and allow swell trains more than 30° off the local wind direction ϕ to keep their identity. For swell trains within 30° of the local wind direction, the wind-sea energy and the potential swell energy values are compared to determine if the swell energy is contained in the local sea or needs to be calculated by the swell energy scanning technique. Additional details on the GONO model are given in a report by Bruinsma et al. (1980).

The GONO is used in an operational mode over the North Sea and uses a rectangular grid as shown in Figure 5.8. Some of the results from the operational testing of the GONO model are presented in Chapter 7.

The GONO model is close in spirit to the models like HYPA (described earlier) or TOHOKU (see The SWAMP Group, 1985) all of which can be classified as Coupled Hybrid (CH) models. A major inadequacy in the formulation of CH models has been the treatment of wind-sea ←→ swell transition regime where the nonlinear energy redistribution is neither negligible nor dominant. Such transition regimes arise whenever the wind speed decreases or the wind direction turns. In view of this, a class of models called Coupled Discrete (CD) has been developed recently in which the traditional discrete spectral representation for both the wind-sea and the swell region has been retained. Some of these models are reviewed in the following section.

5.4 Coupled Discrete Wave Models

The principal distinction between the CH and the CD models lies in the division between the discrete and the parametric representation. In CH models, the entire wind-sea spectrum is treated parametrically and the discrete representation is limited to swell

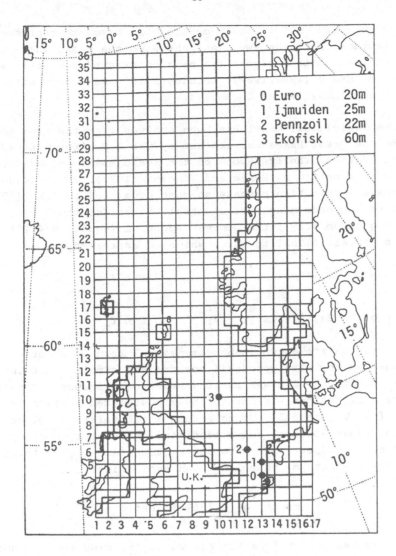

Figure 5.8: The rectangular grid for the GONO model covering the North Sea and vicinity; also shown are waverider locations with the corresponding water depths (from Janssen et al. 1984, Journal of Geophysical Research, Vol. 89; Copyright by the American Geophysical Union)

components only (ex. GONO). In CD models, the discrete representation normally includes both the swell and the wind-sea peak, while the tail-end of the spectrum beyond the peak is treated parametrically. The models of Resio (1981,1987), Golding(1983) and Greenwood et al (1985) can be classified as CD models.

The Resio(1981) model has been evolved from the wave modelling program on the Great Lakes carried out by the U.S. Army Waterways Experiment Station. The framework of the Resio model is based on an earlier model reported by Resio and Vincent (1977). Resio reanalyzed the wave-wave interaction process as conceptualized by Webb (1978) and examined the nondimensional spectral shapes for the P-M spectrum (equation 4.19), the JONSWAP spectrum (equation 4.20) and the Kitaigorodskii (1962) spectrum which is expressed as;

$$
E(f) \left\{ =
\begin{array}{ll}
\alpha \ g^2 f^{-5} (2\pi)^{-4} & f \geq f_m \\[2ex]
\alpha \ g^2 f^{-5} (2\pi)^{-4} \cdot \exp[1 - (\frac{f_m}{f})^4] & f < f_m
\end{array}
\right.
\tag{5.25}
$$

The three spectra are shown in nondimensional form in Figure 5.9. The JONSWAP and the Kitaigorodskii spectra are similar in terms of total energy with the JONSWAP spectrum showing evidence of overshoot-undershoot oscillations similar to those postulated by Barnett and Wilkerson (see Figure 5.3). Using a functional relation between α, the equilibrium range constant and a nondimensional fetch, Resio para-meterized the nonlinear wave-wave interaction term $S_{n\ell}$ in terms of the Kitaigorodskii spectrum; this led to an expression:

$$
S_{n\ell} = \text{const.} \ g^6 (\frac{U}{f_m})^2 (\frac{f}{f_m})^3 \exp[1 - (\frac{f_m}{f})^4]
\tag{5.26}
$$

The above expression involves the square of the wind speed U and this makes the expression comparable to the wind stress on the water surface; this form of parameterization of the nonlinear interaction term is fundamentally different than that proposed by Barnett(1968) or Ewing(1971) in which α is assumed constant. Resio further neglects the linear growth term A (see eg. 5.3) so that the net source function is given by;

$$
S = S_{n\ell} + B \ G(\theta - \phi)
\tag{5.27}
$$

Here B is the exponential growth term and $G(\theta-\phi)$ represents a direc-tional spreading function similar to the one used in the SOWM (see eq. 5.6). The Resio model assumes that waves are always present initially so that the wave-wave interaction term becomes operative and generates more waves; further, a local nonpropagating parametric model with a

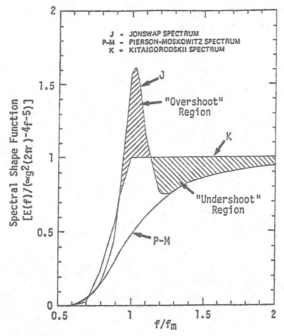

Figure 5.9: A comaparison of nondimensionalized shapes of spectra
proposed by Hasselmann et al (J-JONSWAP), Kitaigorodskii (K) and
Pierson-Moskowitz (P-M) [from Resio, 1981; Copyright by American
Meteorological Society]

high frequency cut-off is incorporated so as to remove excess waves.
The boundary between the parametric and discrete spectral domains of
the model is maintained at a fixed point and energy in each domain is
conserved independently. In a growing wind-sea, energy is initiated
in the parametric region; the rate of change of f_m is calculated
following Hasselmann et al(1976) and energy exchange into the discrete
spectral region is used to trigger the growth in that domain. The wave
growth in the wind-sea region is automatically halted when the nondimensional peak frequency $\nu(= f_m U/g)$ attains a value of 0.13. The
model is said to have reached a fully developed stage at this point
when the growth rate rapidly decreases and the nonlinear interaction
terms distribute energy to lower frequencies; energy residing in the
high frequency part of the spectrum is eventually allowed to dissipate. Further, when the nondimensional peak frequency $\nu < 0.13$, waves
are considered as swells and the spectral components are assumed to
become instantaneously uncoupled and a discrete propagating scheme is
used to advect the energy at the group velocity of each individual
frequency. A swell decay through a wave-wave interaction term appropriate for a swell spectrum is also applied. For numerical integration
of the governing equations, a modified Lax-Wendroff numerical integration scheme (Gadd, 1978) is used for both sea and swell regions. For

model calculations, a special spherical orthogonal grid (see Figure 5.10) has been designed by Resio.

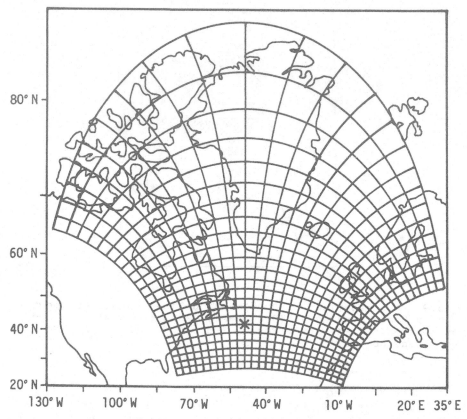

Figure 5.10: **The spherical orthogonal grid for the north Atlantic ocean used in the Resio model (source: Dr. D. Resio)**

The Resio model as described above was tested in a hindcast mode for selected storm cases in the Canadian Atlantic (see Resio, 1982). In a recent study, Resio(1987) has extended his analysis to develop a theoretical framework for the characteristic form of the equilibrium spectra in waters of arbitrary depth. A necessary consequence of the spectral equilibrium range formulation is that a strong constant flux of wave energy exists through this region of the spectrum when an equilibrium spectral form is maintained; the strong energy flux is towards high frequencies where it is lost due to wave breaking. Based on this framework, Resio has developed a spectral wave model applicable to waters of arbitrary depths; additional details of this model together with relevant results are presented in Chapters 6 and 7.

Another model belonging to the CD class is the depth-dependent model developed by Golding(1983) at the British Meteorological Office, Bracknell, U.K. This model which is identified in literature as the BMO model uses the basic energy balance equation (5.1); however, in the presence of variable depth, both the group velocity $\bar{c}_g$ and the direction of propagation θ become functions of time. Following (5.1), the total derivative $\frac{dE}{dt}$ can be expressed as;

$$\frac{dE}{dt} = \frac{\partial E}{\partial t} + \nabla \cdot (\bar{c}_g E) + \frac{\partial}{\partial \theta} (E \frac{d\theta}{dt}) = S (f, \theta, \bar{x}, t) \tag{5.28}$$

write $\frac{d\theta}{dt} = \frac{\partial \theta}{\partial t} + \bar{c}_g \cdot \nabla \theta = \bar{c}_g \cdot \nabla \theta$, since $\frac{\partial \theta}{\partial t} = 0$.

the energy balance equation (5.28) is re-written as

$$\frac{\partial E}{\partial t} = - \nabla \cdot (\bar{c}_g E) - \frac{\partial}{\partial \theta} \{(\bar{c}_g \cdot \nabla \theta)\} + S_{in} + S_{ds} + S_{n\ell} \tag{5.29}$$

$\qquad\qquad\quad\downarrow\qquad\qquad\qquad\quad\downarrow\qquad\qquad\downarrow\qquad\qquad\quad\downarrow$

$\qquad$ propagation $\qquad\quad$ refraction $\quad$ growth + decay $\quad$ nonlinear
$\qquad\qquad\qquad\qquad\qquad\qquad\qquad\qquad\qquad\qquad\qquad\qquad\qquad$ interaction

Equation (5.29) is the basic equation of the BMO's depth-dependent spectral wave model. The various terms on the right side of this equation are expressed as follows:

1. <u>Propagation</u>: $\frac{\partial E}{\partial t} = - \nabla \cdot (\bar{c}_g E)$. This equation is integrated using a modified Lax-Wendroff integration scheme. While solving this equation, c_g the group velocity is considered depth-dependent and is expressed as;

$$c_g = \frac{1}{2} \{1 + \frac{2kh}{\sinh(2kh)}\} \sqrt{\frac{g}{k} \tanh (kh)}$$

Here the wave number k and the water depth h are related by the expression $\sigma^2 = (2\pi f)^2 = gk \tanh kh$. Values of c_g are precomputed for each frequency in the model at intervals of 2 m depth.

2. <u>Refraction</u>: The term $\bar{c}_g \cdot \nabla \theta$ can be expressed as

$$\bar{c}_g \cdot \nabla \theta = - \frac{|\bar{c}_g|}{k} (\frac{\partial h}{\partial x} \sin \theta - \frac{\partial h}{\partial y} \cos \theta) (\frac{-k^2 \text{sech}^2 kh}{\tanh kh + kh \, \text{sech}^2 kh}) \tag{5.30}$$

The first term on the right side of (5.30) is calculated using centered differences for the derivatives of h; the remaining terms are pre-calculated as a function of h for each frequency used in the model.

3. <u>Growth + decay</u>: The energy input S_{in} is represented by linear and exponential terms of the form;

S_{in} = A + BE; this is similar to the growth equation (5.3) used in the PTB model. Field studies by Snyder et al. (1981) suggest that the linear term becomes unimportant soon after initiation of wave growth; hence the following form for A was chosen by Golding:

$$A \begin{cases} = \dfrac{6 \cdot 10^{-8}}{2\pi f_m} \ U^2 \cos^2(\theta - \phi) & \text{for } f = f_m; \quad (\theta - \phi) < 90° \\ = 0 & \text{otherwise} \end{cases}$$

Here U is the wind speed and ϕ is the wind direction. The exponential term B is given by;

$$B \begin{cases} = 6 \cdot 10^{-2} \ (2\pi f \dfrac{\rho_a}{\rho_w}) [\dfrac{U\cos(\theta - \phi)}{c} - 1] & \text{for } \dfrac{U\cos(\theta - \phi)}{c} > 1 \\ = 0 & \text{otherwise} \end{cases}$$

For dissipation, the whitecapping mechanism of Hasselmann (1974) is assumed and this gives S_{ds} = const. $f^2 \bar{E}^{0 \cdot 2 5} E_{ij}$; here $\bar{E}$ is the total spectral energy, E_{ij} is the spectral energy of the ij spectral element and the value of the constant is so adjusted as to produce a duration-limited growth curve based on the JONSWAP data. Further, a quadratic bottom friction law is assumed and an approximate form suggested by Collins (1972) is used to modify the above expression for S_{ds}.

4. <u>Nonlinear interaction</u>: The formulation of this term is based on the parametric wave model of Hasselmann et al. (1976). Here the wind-sea spectrum is separated from the full spectrum at each time step. The energy calculation for swell portion is done by the three preceding steps as discussed above. For the wind-sea spectral region, the three steps as outlined above yield the total energy. The spectral shape appropriate to this energy is obtained by first using the diagnostic relations for the shape parameters (f_m, γ, σ) of the JONSWAP spectrum and then evaluating the shape equation over the appropriate range of parameters. The nonlinear transfer are thus defined implicitly as those required to return the spectrum to this shape. The following shape equation is used:

$$E(f,\theta) = F(f) \exp\{-1.25 \ (\frac{f_m}{f})^4 + \ell n \ \gamma \cdot \exp(\frac{f-f_m}{\sigma f_m})^2\} \ G(f,\theta) \qquad (5.31)$$

Here $\int_\theta G(f,\theta) = 1$ and $\iint E(f,\theta)d\theta df = \bar{E}$, the total energy.

The functions F and G can take various forms. Based on Phillips' (1958) equilibrium range in the wind-wave spectrum, the form for F can be expressed as $F(f) \alpha f^{-5}$; with this form, equation (5.31) becomes the normal JONSWAP spectrum given by (4.20). Alternative forms of F for shallow water spectrum have been suggested by Kitaigorodskii et al (1975); a recent study by Kitaigorodskii (1983) suggests that the equilibrium form of the wind-wave spectrum may have two asymptotic regimes, one proportional to f^{-4} and the other to f^{-5}.

The function G is the angular spreading function and is generally expressed as the square of the cosine of the angle between wind and wave direction $[\cos^2(\theta-\phi)]$.

The separation of the wind-sea spectrum from the full spectrum is done on the following basis:

$$f > 0.8f_m \text{ and } |\theta-\phi| \leq 90°$$

Here, the condition of angular deviation can be easily applied. To apply the frequency constraint, the peak frequency f_m must be known. An iterative procedure is used which starts with the lowest possible frequency given by the P-M spectrum and is then modified using the total wind-sea energy calculated at every time step. The values f_m and γ can be redefined based on the wind-sea energy values and using the peak width $\sigma = 0.08$, the shape equation (5.31) can be recalculated and normalized so that the redefined spectrum has the same energy as the original. The overall effect of this process is to replace the wind-sea part of the spectrum by a JONSWAP spectrum with the same energy. This allows the discrete spectral components of the model to be coupled through the nonlinear interaction process; accordingly, the BMO's model and other models using similar techniques are categorized as CD models.

The BMO model was originally developed on a nested grid consisting of a coarse grid with a 300-km grid spacing (at 60°N) and a nested fine grid with a grid spacing of 50 km. The coarse grid covered the north Atlantic and a adjoining land areas between latitudes 20°N and 75°N and between longitudes 110°W to 20°E, while the fine grid covered the northwestern European continental shelf area. A deep-water version of the model was applied over the coarse grid providing boundary conditions for the fine grid over which the depth-dependent model as described above was applied. Since April 1987, the BMO model has been upgraded and transformed to a global model which uses a grid spacing of 1.5° latitude by 1.875° longitude everywhere over the global oceans, while over European waters, a separate finer grid (0.25° lat. x 0.4° long.) has been designed. The two grid (coarse and fine) of the BMO model are displayed in Figures 5.11 and 5.12 respectively. The model physics has been upgraded to include a modified

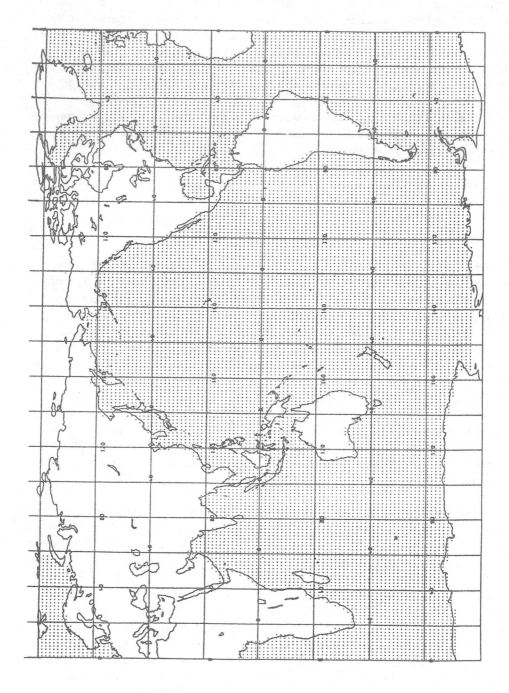

Figure 5.11: The global grid of the BMO operational wave model (Source: Dr. Rachel Stratton, British Meteorological Office)

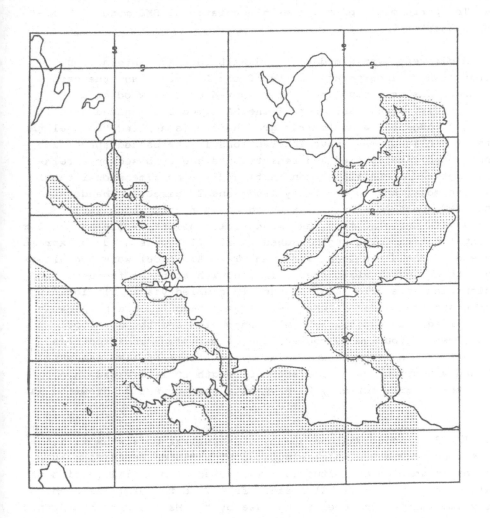

Figure 5.12: The European grid of the BMO operational wave model (Source: Dr. Rachel Stratton, British Meteorological Office)

linear growth term, an improved dissipation term (based on a form
proposed by Komen, Hasselmann and Hasselmann, 1984) and an inclusion
of a directional relaxation term which allows the higher frequency
waves to adjust to wind direction faster than the lower frequency
waves but still in a finite time. Further, the global version of the
model uses a great circle turning algorithm for swell propagation and
an incorporation of an ice edge position based on weekly sea-ice
charts. The performance of the present operational BMO model is discussed in Chapter 7.

A discrete/hybrid model called SAIL (Sea Air Interaction
Laboratory) which incorporates parts of the PTB model and the parametric model has been developed and reported by Greenwood et al.
(1985). The directional spectrum of the SAIL model is banded and all
bands are propagated; a parametric growth form is applied to model the
effects of nonlinear wave-wave interactions in the banded space. The
frequency bands are coupled and redistribute energy based on a relationship between α (Phillips' constant) and the nondimensional total
energy. The model has 15 frequency bands and 24 direction bands and
uses the P-M spectrum with a peak enhancement factor which varies
inversely with the nondimensional mean square height, while an angular
spreading function following Mitsuasu et al. (1975) is used to express
the spread of the directional spectrum. The SAIL model works on alter-
nate growth and propagation steps. In a growth step the frequency and
direction bins at a grid point are weakly coupled via a relationship
between α and the nondimensional mean square height. In a propagation
step, the frequency bands are totally uncoupled and the direction
bands are weakly coupled by convergence of meridians on a spherical
earth. Further, in order to approximate the effect of simultaneous
growth and advection, each time step of length Δt is divided into
three parts: 1. grow for $\frac{1}{2}\Delta t$, 2. propagate for Δt, and 3. grow for $\frac{1}{2}\Delta t$
again.

The SAIL model was one of the ten spectral models used in the
SWAMP testing program in which simulated wind fields pertaining to a
variety of meteorological situations were used. A modified version of
the SAIL model covering global oceans (from 75°N to 70°S) has been
recently implemented for operational use by the Marine Products Branch
of the National Oceanographic and Atmospheric Administration (NOAA) in
Washington, U.S.A. (see Esteva and Chin,1987).

5.5   Third Generation Wave Models

The models described in sections 5.3 and 5.4 may be called the
second generation spectral wave models since these models include
a parameterized version of the wave-wave interaction process which

is essentially absent in the first generation spectral wave models described in section 5.2. A major thrust towards development of the third generation wave model was provided in a paper by Hasselmann and Hasselmann (1985a) in which a new method for efficient computations of the exact nonlinear transfer integral (eq. 3.10) was demonstrated. With increasing availability of supercomputers, the third generation wave models have been developed as a research tool and are being tested in a semi-operational mode at present.

The model EXACT-NL described by Hasselmann and Hasselmann (1985b) was an attempt to include the exact form of the nonlinear transfer term $S_{n\ell}$ at least for simple cases in which only one integration variable occurs; the basic governing equation (5.1) was used with S_{in} the wind input term prescribed following Snyder et al (1981) and S_{ds} the dissipation term based on an empirically derived expression. The EXACT-NL model was one of the ten spectral wave models used in the SWAMP study in which three cases were investigated using the model; fetch- and duration- limited wave growth for a uniform wind field case and a sudden 90° change in the wind direction of an otherwise stationary, uniform wind field. For the fetch-or duration-limited wave growth cases, a strictly stationary fully developed equilibrium spectrum is never established in the EXACT-NL model; this is because the nonlinear source term continues to transfer energy very slowly from the peak region of the spectrum to the lower frequencies while the energy gained by the low frequencies cannot be balanced by the dissipation and the input source terms used in the model. Appropriate forms of the input and dissipation terms in the vicinity of the peak frequency appears to be important in controlling the energetics of the EXACT-NL model.

The SWAMP study further revealed a number of basic shortcomings of the first generation as well as of the second generation models. These deficiences were associated with inadequate parameterization of the nonlinear transfer processes and became particularly evident in rapidly changing wind fields and in the interaction of wind-sea with swell. When the SWAMP models were driven by identical hurricane wind fields, the various models generated maximum significant wave heights which ranged from 8 to 25 m (Komen, 1987). Similar other discrepancies in the SWAMP study have led to the formation of a WAM (WAve Modelling) Group consisting of international wave modellers (primarily from European countries). The WAM Group has developed a third-generation model which uses the discrete interaction approximation of Hasselmann et al (1985); this approximation retains the same cubic operator structure as in the original Boltzmann integral. The basic energy transport equation for the WAM model is developed for a spherical latitude-longitude (ϕ, λ) coordinate system. For the two-

dimensional ocean wave spectrum $E(f,\theta,\phi,\lambda,t)$, the transport equation
can be expressed as,

$$\frac{\partial E}{\partial t} + (\cos\phi)^{-1}\frac{\partial}{\partial\phi}(\dot{\phi}\cos\phi E) + \frac{\partial}{\partial\lambda}(\dot{\lambda}E) + \frac{\partial}{\partial\theta}(\dot{\theta}E) = S \qquad (5.33)$$

Here S is the net source function describing the change of energy of
a propagating wave group and the rates of changes of ϕ(latitude),
λ(longitude) and θ(direction, measured clockwise relative to true
north) are given by

$$\dot{\phi} = \frac{d\phi}{dt} = c_g R^{-1}\cos\theta$$

$$\dot{\lambda} = \frac{d\lambda}{dt} = c_g\sin\theta(R\cos\phi)^{-1} \qquad (5.34)$$

$$\dot{\theta} = \frac{d\theta}{dt} = c_g\sin\theta\cdot\tan\phi\cdot R^{-1}$$

Equations(5.34) represent the rates of changes of the position and
propagation direction of a wave packet travelling along a great circle
path. In these equations, c_g refers to the deep-water group velocity
($c_g = g/4\pi f$) and R is the radius of the earth. The expressions for
various source functions on the right hand side of (5.33) are developed based on earlier discussions.

The WAM model has been developed on a global domain as well as
on a regional domain. For the global domain a 3°x3° grid has been
designed, while for the regional domains (North Atlantic and the Gulf
of Mexico) a latitude-longitude grid of ¼°x½° has been designed. At
present, the WAM model has been implemented on the CRAY computer of
the European Centre for Medium-Range Weather Forecasts (ECMWF) in
Reading, United Kingdom and has been tested on selected case studies
using wind fields from the ECMWF weather prediction model as well as
from a fine-mesh model of the British Meteorological Office. A few
results from these studies have been reported by Komen and Zambresky
(1987) while the full details of the WAM model are given in a comprehensive paper by the WAMDI Group (1988).

In summary, a majority of the present operational wave prediction models belong to the CH or CD class, although a number of first
generation wave models belonging to the DP class are still in operational use, for ex. the GSOWM of the U.S. Navy. The present trend
appears to be towards refinement of existing CH or CD models both of
which are catagorized as second generation models. With increasing
computer power becoming available in real-time operational environment, it is envisaged that the third generation model WAM or its
variant may become operational at meteorological and oceanographic
centres of many countries.

REFERENCES

Baer, L. 1962: An experiment in numerical forecasting of deep water ocean waves. Report No. LMSC-801296, Lockheed California Company, U.S.A.,

Barnett T.P., 1968: Loc. cit. (Chapter 3)

" and J.C. Wilkerson, 1967: On the generation of wind waves as inferred from air-borne measurements of fetch-limited spectra. J. Marine Research, 25, 292-328.

Bruinsma, J., P.A.E.M. Janssen, G.J. Komen, H.H. Peeck, M.J.M. Saraber and W.J.P. de Voogt, 1980: Description of the KNMI operational wave forecast model GONO. Scientific Report 80-8, Royal Netherlands Meteorological Institute, De Bilt, The Netherlands.

Burling, 1959. The spectrum of waves at short fetches. Deut. Hydrogr. Zeitschriff 12, 45-64, 96-117

Cardone, V.J., W.J. Pierson and E.G. Ward, 1976. Hindcasting the directional spectra of hurricane generated winds. J. Petroleum Technology, 28, 385-394.

Clancy R.M., J.E. Kaitala and L.F. Zambresky, 1986: The Fleet Numerical Oceanography Centre Global Spectral Ocean Wave Model. Bull. Amer. Meteor. Society, 67, No. 5, 498-512.

Collins, J.I., 1972: Prediction of shallow water spectra J. Geophysical Research, 77, 2693-2707

Esteva D. and H. Chin, 1987; Development of a global scale ocean wave forecasting model for marine guidance. Proc. Int'l workshop on wave hindcasting and forecasting, Halifax, Nova Scotia, Sept. 23-26, 1986. Env. Studies Revolving Fund, Report Series No. 065, Ottawa, p.99 (Abstract only)

Ewing, J.A., 1971: A numerical wave prediction method for the north Atlantic ocean. Deut. Hydrogr. Zeitschriff, 24, 241-261

" , T.J. Weare and B.A. Worthington, 1979: A hindcast study of extreme wave conditions in the north sea. J. Geophysical Research, 84, 5739-5747.

Gadd, A.J., 1978: A numerical advection scheme with small phase speed errors. Q.J. Royal Meteorological Society, 104, 583-594.

Gelci, R., H. Cazale and J. Vassal, 1957: Prévision de la Houle, La Methode des Densités Spectro-Angulairies. Bulletin d'information, Comité Central d'Oceanographie et d'Etude des Cotes, 8, 416-435.

" E. Devillaz and P. Chavy, 1964: Evolution de l'Etat de la Mer, Calcul numerique des advection. Note de l'Etablissment d'Etude et de Recherches Météorolgique, No. 166, Paris, France

" and P. Chavy, 1978: Seven years of routine numerical wave prediction by DSA-5 model. Turbulent Fluxes Through the Sea Surface, Wave Dynamics and Prediction, (Ed. A. Favre and K. Hasselmann), Plenum, 565-591.

Golding, B., 1983: A wave prediction system for real-time sea-state forecasting. Q. J. Royal Met. Soc., 109, 393-416

Günther, H. and W. Rosenthal, 1985: The Hybrid Parametrical (HYPA) wave model. Ch. 20, Ocean wave Modelling. The SWAMP Group, Plenum, 211-214

"   " and Dunkel, 1981: The response of surface gravity waves to changing wind direction. J. Physical Oceanography, 11, 718-728

"   ", T.J. Weare, B.A. Worthington, K. Hasselmann and J. Ewing, 1979: A Hybrid-Parametrical wave prediction model. J. Geophysical Research, 84, 5727-5737.

Hasselmann, K., 1962: Loc. cit. (Ch. 3)

Hasselmann, K., 1963: Loc. cit. (Ch. 3)

Hasselmann, K. et al., 1973: Loc. cit. (Ch. 3)

Hasselmann, K., B. Ross, P. Müller and W. Sell, 1976: A parametric wave prediction model. J. Physical Oceanography, 6, 200-228

Hasselmann, K., 1974: Loc. cit. (Ch. 3)

Hasselmann, S. and K. Hasselmann, 1985a: Computations and parameterizations of the nonlinear energy transfer in a gravity wave spectrum. Part I. A new method for efficient calculations of the exact nonlinear transfer integral. J. Physical Oceanography, 15, 1369-1377.

"   " 1985b: The wave mdoel EXACT-NL. ch. 24, Ocean wave modelling, The SWAMP Group, Plenum 249-251.

"   ", J.H. Allender and T.P. Barnett, 1985: Computations and parameterizations of the nonlinear energy transfer in a gravity wave spectrum. Part II. Parameterization of the nonlinear energy transfer for application in wave models. J. Physical Oceanography, 15, 1378-1391.

Hydraulics Research Station (HRS), 1977: Numerical wave climate study for the North Sea (NORSWAM). Report No. Ex 775, Hydraulics Research Station, Wallingford, U.K., 17 pp+ Figures and Tables.

Janssen, P.A.E.M., G.J. Komen and W.J.P. De Voogt, 1984: An operational coupled hybrid wave prediction model. J. Geophysical Research, 89, C3, 3635-3654.

Kitaigorodskii, S., 1962: Application of the theory of similarity to the analysis of wind-generated wave motion as a stochastic process. Bull. Aca. Sci. USSR Ser. Geophy. No. 1, Vol. 1, 105-117.

"   1983: On the theory of equilibrium range in the spectrum of wind generated gravity waves. J. Pysical Oceanography, 13, 816-827.

"   V.P. Krasitskii and M.M. Zeslavskii, 1975: On Phillips' theory of equilibrium range in the spectra of wind generated waves. J. Physical Oceanography, 5, 410-420.

Komen, G., 1987: Recent results with a third generation ocean-wave model. Johns Hopkins APL Technical Digest, Vol. 8, No. 1, 37-41.

"   , S. Hasselmann and K. Hasselmann, 1984: On the existence of a fully developed wind-sea spectrum. J. Physical Oceanography, 14, 1271-1285.

"   and L. Zambresky, 1987: A Third Generational Ocean wave model. Proc. Int'l Workshop on wave hindcasting and forecasting, Halifax, Canada, Sept. 23-26, 1986. Env. Studies Revolving Fund, Report Series No. 065, Ottawa, 233-242.

Miles, J.W., 1957: Loc. cit. (ch. 3)

Mitsuasu, H., 1968: On the growth of the spectrum of wind-generated waves (I). Rep. Res. Inst. Appl. Mech., Kyushu University, Japan, 16, 459-482.

Mitsuasu, H., F. Tasai, T. Suhara, S. Mizuno, M. Ohkusu, T. Honda and K. Rikiishi, 1975: Observations of directional spectrum of ocean waves using a cloverleaf buoy. J. Physical Oceanography, 750-760

Mitsuasu, H., 1982: Wind wave problems in engineering. Engineering Meteorology (Ed. E.J. Plate), Elsevier, ch. 15, 683-729.

Phillips, O.M., 1957: Loc. cit. (Ch. 3)

Pierson, W.J., 1982: The Spectral Ocean Wave Model (SOWM), A northern hemisphere computer model for specifying and forecasting ocean wave spectra. Final Report No. DTNSRDC-82/11, U.S. Naval Oceanography Command Detachment, Asheville, North Carolina, 201 pp.

"   and L. Moskowitz, 1964: Loc. cit. (Ch. 4)

"   , L.J. Tick and L. Baer, 1966: Computer based procedure for preparing global wave forecasts and wind field analysis capable of using wave data obtained by a space craft. Proc. Sixth Naval Hydrodynamics Symp., Washington, 499-533.

Resio, D.T., 1981: The estimation of wind-wave generation in a discrete spectral model. J. Physical Oceanography, 11, 510-525.

"   1982: Assessment of wave hindcast methodologies in the Scotian Shelf, Grand Banks and Labrador Sea areas. Canadian Contractors Report, Hydrography and Ocean Science, 4, Ottawa, 128 pp. July 1982.

"   1987: Shallow-water waves. I. Theory. J. Waterways Port, Coastal and Ocean Engineering, 113, 264-281

"   and C.L. Vincent, 1977: A numerical hindcast model for wave spectra on water bodies with irregular shoreline geometry. Rept. 1: Test of nondimensional growth rates. Misc. Paper H-77-9, Hydraulics Lab. U.S. Army Engineer Waterways Experiment Station, 53 pp.

Synder, R.L., et al., 1981: Loc. cit. (Ch. 3)

The SWAMP Group, 1985: Ocean Wave Modelling, Plenum, 256 pp.

The WAMDI Group, 1988: The WAM model - A third generation ocean wave prediction model. J. Physical Oceanography, 18, 1775-1810.

Weare T.J. and B. Worthington, 1978: A numerical model hindcast of severe wave conditions for the North Sea. Turbulent Fluxes Through the Sea Surface, Wave Dynamics and Prediction (A. Favre and K. Hasselmann, Eds.) Plenum, 617-628.

Webb, D.J., 1978: Non-linear transfer between sea waves. Deep-Sea Research, 25, 279-298.

CHAPTER 6
WAVE PREDICTION IN SHALLOW WATER

6.1 General Comments

Prediction of wave conditions in shallow water has received
increased emphasis in both the scientific and engineering communities
in recent years. The scientific interest in the problem of shallow-
water wave prediction has increased because of generally increased
knowledge about wave generation, improved meteorological and oceanographical database and the availability of bigger and better computer
facilities to handle the mathematical and numerical aspects of the
problem. Increased marine activities, particularly in areas of offshore exploration and coastal development has created a need for
improved knowledge of sea-state conditions in the nearshore regions.
As a result, a substantial amount of research and developmental work
on shallow-water wave modelling and prediction has been initiated in
recent years. In the following section, a brief review of early work
on wave analysis and prediction in shallow water is presented; this is
followed by a review of more recent studies, particularly those based
on a spectral approach. The last section discusses inclusion of
shallow-water effects in operational wave analysis and prediction.

6.2 Review of Earlier Work

The problem of wave analysis and prediction in shallow water and
in coastal regions has been recognized for more than fifty years. Two
basic approaches to shallow-water wave modelling are identified at
present; one, in which the entire wave generation area lies in relatively shallow water and the second, in which the waves are generated
in deep water and then propagated into shallow water. The first
approach is important in selected areas such as semi-enclosed basins,
shallow bays or regions where ice cover can restrict the fetch area
to a shallow coastal zone. The second approach is important to all
coastal areas where large storm waves propagate from deep water into
shallow coastal waters. Most of the early work is concerned with the
second approach namely transformation of deep-water waves as they
propagate into coastal waters. As we have seen earlier in Chapter 2,
once the water depth becomes less than about a half of the wavelength,
the speed of the wave crest (the phase speed c) and of the wave energy
(the group wave speed c_g) become depth-dependent. Thus, an initially

straight wave front propagating into water of varying depth will become curved in a manner analogous to the refraction of light waves; this has led to the preparation of first water wave refraction diagrams in which wave rays (lines perpendicular to wave front) were plotted and wave fronts were constructed using Huygen's principle. The wave ray equations given by Munk and Arthur(1952) are expressed as

$$\frac{dx}{ds} = \cos\theta \; ; \quad \frac{dy}{ds} = \sin\theta$$

$$\frac{d\theta}{ds} = \frac{1}{c}[\sin\theta\,\frac{\partial c}{\partial x} - \cos\theta\,\frac{\partial c}{\partial y}]$$

(6.1)

Here ds is an infinitesimal increment along the wave ray, x, y are the two horizontal axes and θ is the angle the ray makes with the x axis. The wave rays represent the direction in which the wave fronts are propagating. In the presence of refraction, the wave rays are curved and in general may tend to diverge or converge. Over a submarine canyon, the wave fronts will always tend to diverge (see Figure 6.1a) while over a submarine ridge the wave fronts will always tend to converge (see Figure 6.1b); for a specialized case of a uniformly sloping beach with depth contours parallel to the coast line, wave fronts propagating at an angle to the coast line will become almost parallel as the waves approach the coast line (see Figure 6.1c).

In order to calculate the wave energy changes along wave rays, it is generally assumed that wave energy between two rays remains constant as the wave front progresses which means that there is no dispersion of energy laterally along the front, no reflection of energy from the rising bottom and no loss of energy by any other processes. With these assumptions, we can write

$$E\,c_g\,b = E^d c_g^d b^d$$

(6.2)

Here E is the wave energy per unit surface area, b is the distance between any selected pair of wave rays and the superscript d refers to deep-water conditions. From (6.2), the proportional change in wave height H from deep water to any shallow water point can be expressed as

$$\frac{H}{H^d} = (\frac{E}{E^d})^{\frac{1}{2}} = (\frac{c_g^d}{c_g})^{\frac{1}{2}}(\frac{b^d}{b})^{\frac{1}{2}} = K_s K_r$$

(6.3)

In (6.3), K_s is called the shoaling factor and K_r is called the refraction factor. If the depth contours are parallel to the coast line as in Figure 6.1c, the refraction factor K_r can be expressed as

$K_r = \sqrt{\cos\theta^d/\cos\theta}$ where θ^d is the angle the wave front makes with the

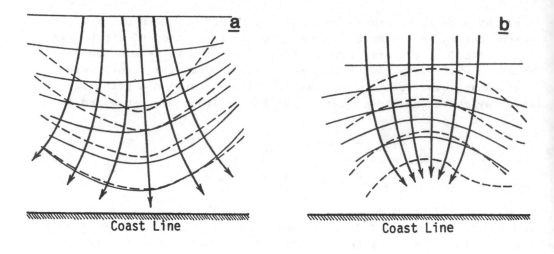

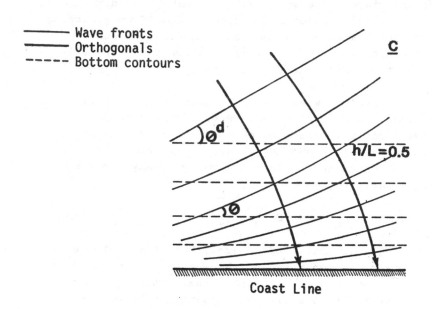

Figure 6.1: <u>a</u>. Diverging wave rays (orthogonals) over a submarine canyon <u>b</u>. Converging wave rays over a submarine ridge <u>c</u>. Refraction of wave rays approaching a gently sloping coast line at an angle θ.

bottom contours in deep water (see Fig. 6.1c) while θ is the corresponding angle at a shallow water location.

Equation (6.3) is the basic equation which has been used by
several investigators (ex. Johnson et al. 1948; Griswold, 1962;
Wilson, 1966; Aranuvachapun, 1977) to develop algorithms for plotting
surface wave rays over different coastal regions. The shoaling factor
K_s and the refraction factor K_r both depend upon the depth and
produce opposite effects on the wave height, in general. Refraction
commonly tends to increase the length of the wave crest and thus
reduce the wave height while the shoaling factor tends to increase the
wave height except in a relatively unimportant range of depth where
the waves first feel the bottom. Using appropriate equations based on
Snell's law of refraction, Johnson et al(1948) have constructed a
nomogram which gives changes in wave direction and height due to
refraction on shores with straight parallel depth contours. In later
studies, Griswold(1963) for example, discusses computer implementation
of refraction diagrams and refraction coefficients by Munk and Arthur
method, while Wilson(1966) gives computer programs for plotting wave
rays.

In using Eq.(6.3) as discussed above, no account is taken of the
energy loss due to bottom friction and due to percolation in the
permeable bed. One of the earliest attempts to quantify the dissipation of wave energy by bottom friction was that of Putnam and Johnson
(1949) who considered the oscillatory motion of waves at the sea
bottom. By suitable numerical examples, Putnam and Johnson showed that
the loss of wave energy due to bottom friction could reduce the wave
height by as much as 30 percent on very flat beaches for wave periods
commonly occurring in the ocean. The classical work of Bretschneider
and Reid(1954) extends the study of Putnam and Johnson and develops
appropriate formulas and graphs for facilitating the computation of
wave height changes resulting from the combined effects of friction,
percolation and refraction together with the direct shoaling effect.
Considering the rate of change of energy in terms of power (P) transmitted across the vertical section bh(where h is the water depth)
between two consecutive wave rays (separated by a distance b), one can
express the ratio of power of transmission from deep to shallow-water
as

$$\frac{H}{H^d} = (\frac{b^d}{b})^{\frac{1}{2}}(\frac{c^d_g}{c_g})^{\frac{1}{2}}(\frac{Pb}{P^d b^d})^{\frac{1}{2}} \quad \rightarrow \quad \text{this can be}$$

expressed as $\qquad \frac{H}{H^d} = K_r K_s K_{pf}$ $\hspace{4cm}$ (6.4)

Equation (6.4) has one additional coefficient than (6.3) namely K_{pf}

which is identified as a friction-percolation coefficient and whose
value is < 1. Bretschneider and Reid have further analyzed eq.(6.4) in
an attempt to obtain a general solution for the wave reduction factor;
their analysis shows that the coefficient K_{pf} is not independent of
wave refraction; consequently, eq.(6.4) is re-written as

$$\frac{H}{H}d = K_s K_{pfr} \tag{6.5}$$

Here the coefficient K_{pfr} represents the combined effects of percolation, friction and refraction. Equation(6.5) is solved in conjunction
with other related equations to obtain wave height changes due to the
effects of friction, percolation and refraction. As an example, Table
6.I which is reproduced from Bretschneider and Reid(1954), shows
values of K_f, K_p and K_{pf} for selected distances measured from the
deep-water location as defined by the criteria $\frac{h}{L}$ = 0.5. The Table
shows that for a flat beach with a slope of 1:300, a 12-second wave
having a wave height of about 1.6m at a location where the ratio of
water depth to wave length is 0.5, will suffer a loss in wave height
of 21 percent due to friction alone as the wave approaches the
theoretical breaking point; the wave height loss due to percolation
for typical values of sand permeability is only about 7 percent over
the same distance. For a steep beach (slope 1:10), the calculations
show that the same wave will suffer a loss in wave height of only 1
percent due to friction and less than 0.5 percent due to percolation,
over the same distance. The wave height changes due to refraction and
direct shoaling effects only, are governed by eq.(6.3); wave refraction causing convergence (Fig. 6.1b) will lead to increase in wave
height while wave refraction causing divergence (Fig. 6.1a) will lead
to decrease in wave height.

Refraction of waves due to currents can be important in shallow
water, particularly in coastal areas where strong tidal and longshore
currents are known to exist. The influence of currents can be included
by replacing the phase speed c in the above equations by the absolute
speed c_a given by

$$c_a = c + U_c \cos\phi \tag{6.6}$$

Here U_c is the current speed and ϕ is the angle between the current
direction and the wave propagation direction. The direction of propagation of energy, according to the classical study of Johnson(1947),
is however not the same as the wave orthogonal direction except in the
special case of co-linear waves and currents; further, the calculation
of wave height or wave energy becomes more difficult when currents are
present since energy is exchanged between the waves and the currents.

TABLE 6.I: Values of K_f (friction factor), K_p (percolation factor) and K_{pf} (friction and percolation factor) for selected distances measured from the deep-water location. Given conditions: f (coefficient of friction for bottom) = 0.01; m (bottom slope) = 1/300; H^d (wave height in deep water) = 1.6 m; T (wave period) = 12 s; p (permeability coefficient) = 9.876 x 10⁻¹¹ m²; ν (kinematic coefficient of viscosity for water) = 9.662 x 10⁻⁷ m² s⁻¹

Percent Distance*	$\frac{h}{T^2}$ (m s⁻²)	K_f	Percent loss of wave height (friction only)	K_p	Percent loss of wave height (percolation only)	K_{pf}	Percent loss of wave height (friction + percolation)
0.0	0.780	1.00	0	1.00	0	1.00	0
20.7	0.624	1.00	0	0.999	0.1	0.999	0.1
51.7	0.390	0.999	0.1	0.996	0.4	0.995	0.5
70.3	0.250	0.996	0.4	0.989	1.1	0.985	1.5
80.5	0.172	0.990	1.0	0.981	1.9	0.971	2.9
90.8	0.094	0.965	3.5	0.965	3.5	0.932	6.8
95.0	0.062	0.931	6.9	0.954	4.6	0.890	11.0
97.1	0.047	0.909	9.1	0.946	5.4	0.862	13.8
100.0	0.023	0.781	21.9	0.927	7.3	0.734	26.6

* Distance measured from point where $\frac{h}{L}$ = 0.5 to theoretical breaking point with no losses (from Bretschneider and Reid, 1954)

Several observational studies have confirmed a significant impact of
currents on wave height and wave energy. The Agulhas current off South
Africa has been blamed for generating abnormally large waves, sufficient to destroy deep-sea vessels (Sanderson, 1974; Sturm, 1974). The
impact of tidal currents on wave height and wave energy has been
reported by Vincent (1979) in the North Sea and more recently by
Lambrakos(1981) in the Strait of Juan de Fuca off British Columbia on
the west coast of Canada. The problem of wave-current interaction has
been studied from a modelling point of view, by a few investigators in
recent years; for example, Hays(1977, 1980) describes an operational
model for the calculation of refractive effects of the current with
particular reference to the Gulf Stream, while Jonsson and Wang(1980)
discuss equations for current-depth refraction based on wave action
and radiation stress concept and give solutions for idealized cases.
The wave-current interaction equations are complex and have not been
routinely used in practical refraction calculations so far.

6.3 Spectral Approach

The studies summarized above have primarily utilized the classical concepts of wave ray geometry and energy conservation along a wave
ray. With the development of a wave spectrum concept, several investigators have considered a spectral approach to study the shallow-water
effects. Longuet-Higgins(1966) was among the first to study the evolution of spectra in shallow water and showed that the energy density,
if defined per unit area of wave number space, remains constant along
a wave ray. Karlsson(1969) interpreted the result of Longuet-Higgins
as equivalent to the energy conservation equation (6.2) where E is
defined as the energy density in the (f, θ) space. Collins(1972)
developed a numerical model of spectral changes in shallow water and
included, in his model wave generation, a simplified form of bottom
friction dissipation and an equilibrium spectrum form of wave breaking. Using the ray technique of Munk and Arthur(1952), Collins calculated the shallow water spectra for linear parallel bottom contours as
well as for a fully three-dimensional bathymetry. Collins applied his
model to a set of wave data collected off the Florida coast during the
passage of hurricane Betsy(9-10 September 1965) in the Gulf of Mexico
and concluded that a simple topography represented by parallel bottom
contours could very well simulate the effect of bottom friction.

Several studies on computations of shallow-water wave spectra
have been reported since Collins work. Shiau and Wang(1977) considered
the energy conservation equation (6.2) expressed in spectral form and
assuming straight and parallel bottom contours to represent bathymetry, solved the energy conservation equation to obtain spectral
transformation from deep water to shallow water. Wang and Yang(1981)

upgraded the model of Shiau and Wang to include bottom friction and
applied the same to the wave spectra measured in the nearshore region
extending approximately 900 m from the shore at the Island of Sylt in
the North Sea, the same location as the JONSWAP field experiment (see
Fig. 5.4). No wave generation effects were included in the model but
an empirical wave breaking criterion was used to limit the growth of
the wave spectrum as it approached the shoreline. The model calculations were done to obtain wave spectral transformation from deep-water
sites to shallow-water sites. In Figure 6.2 are shown the wave transformation results at location S_2(225 m from the shore) obtained by
Wang and Yang using field data input at location S_1(940 m from the
shore); similar other model calculations were made to obtain wave
transformations into nearshore locations closer to the beach. In
general, the model of Wang and Yang produced acceptable but highly
variable results, with inclusion of bottom friction showing a slight
improvement. In another study, Cavaleri and Rizzoli (1981) used
similar techniques as those of Collins(1972) and calculated wave
spectra for the Adriatic Sea and the Tyrrhenian Sea. They included in
their model, wave generation, refraction and shoaling explicitly; the
nonlinear dissipative forces were introduced through a suitable parameterization. The model was used in a hindcast mode to generate one-
dimensional spectrum and significant wave height at a target point (in
16 m water depth) for selected storm periods in the Adriatic Sea. An
evaluation of model results with observed data gave the Root Mean
Square (RMS) error for significant wave height between 0.15 and 0.5 m.
According to the authors (Cavaleri and Rizzoli), the model predictions
appeared sensitive to the wind field specification, a problem inherent
with all wave hindcasting and wave forecasting models. The study by
Shemdin et al(1980) represents what is probably the most complete
calculations of wave spectral transformation in shallow water; these
include wave generation, bottom friction, wave-wave interaction,
percolation and wave-induced bottom motion. The model calculations
were performed over an area north of the JONSWAP line (see Fig. 5.4b)
in the North Sea where the bottom sediment was known to consist of
medium to fine sand grain; such a sediment tends to make bottom friction more important than percolation. The wave energy transformations
between 17 m and 13 m depth over a propagation distance of about 18 km
were calculated by the model and are shown in Figure 6.3. The measured
spectrum, in this case has two dominant frequency peaks as depicted by
the solid curve in Fig. 6.3. The computed energy changes between two
JONSWAP stations as displayed in the lower panel of Fig. 6.3 shows
further that the most dominant wave energy transformation is due to
frictional dissipation which is followed by refraction and shoaling
effects; the dissipation due to percolation was found to be small
compared to bottom friction. The agreement between measured and computed spectrum is generally good except at high frequencies where

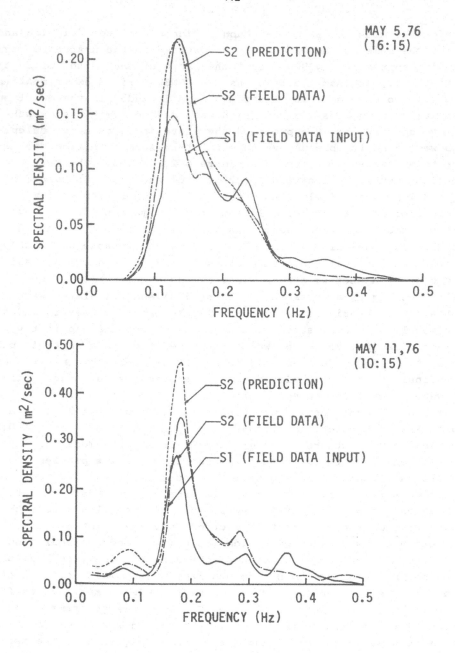

Figure 6.2: Comparison of measured and computed spectral transformation from deep to shallow water. Top: Data of 5 May 1976, 1615 hr Bottom: Data of 11 May 1976, 1015 hr (from Wang and Yang, 1981)

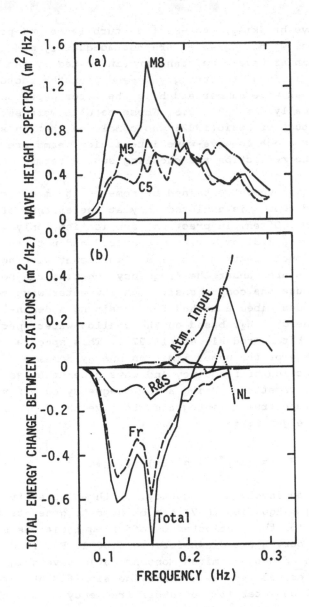

Figure 6.3: Wave transformation for finite depth water for 23 August 1975, 2000 GMT, north of the JONSWAP array. Upper Panel: M8: Measured spectrum at station 8; M5: Measured spectrum at station 5; C5: Computed spectrum at station 5 using measured spectrum at station 8 as input. Lower Panel: Total computed changes between station 8 and 5. Fr: Frictional dissipation; R&S: Refraction and Shoaling; NL: Nonlinear wave-wave transfer (from Shemdin et al. 1980, Journal of Geophysical Research, Vol. 89; Copyright by American Geophysical Union)

processes like wave breaking, near-surface turbulence and possibly strong interactions with long waves are not adequately modelled. Further, the nonlinear wave-wave transfer (indicated by the curve labelled NL) term is found to strongly depend upon the factor $k_p h$ where k_p is the peak wave number and h is the water depth and increases dramatically for $k_p h \leq 0.5$. These results were substantiated in a later study of Resio(1982) who concluded that in shallow water, energy losses due to wave-wave interaction terms are considerably larger than the refraction and shoaling source terms.

The modest improvement obtained by some of the model calculations as described above, is achieved only at the expense of substantial computing time. Often, in practical application, only the wave height information in shallow water is required and not the detailed one or two-dimensional spectrum. Thus, a significant wave height, proportional to the area under the frequency spectrum can be estimated without having to use the comprehensive shallow-water wave models. Vincent(1982) has described a method for obtaining a depth-limited significant wave height, H_ℓ, based on the shallow-water spectral form described by Kitaigorodskii et al(1975). This spectral form provides an estimate of the upper limit on energy density in water depth h as a function of frequency and wave generation as expressed in the Phillips' parameter α. Using a low frequency cutoff value f_c, the shallow-water spectrum is integrated to give the depth-limited significant wave height H_ℓ as

$$H_\ell = 4[\int_{f_c}^{\infty} \alpha \, g^2 (2\pi)^{-4} \phi(\omega_h) df]^{\frac{1}{2}} \tag{6.7}$$

Here ϕ is a function involving frequency, depth and gravity and $\omega_h = \omega(h/g)^{\frac{1}{2}} = 2\pi f(h/g)^{\frac{1}{2}}$. Equation (6.7) can be used to generate typical curves for H_ℓ for three selected cutoff frequencies as shown in Figure 6.4; these curves are drawn for $\alpha = 0.0081$. For ready comparison, the variation of depth-limited monochromatic wave height H_d is also shown. These curves can readily provide significant wave height estimates for shallow water if the cutoff frequency f_c and the Phillips' parameter α can be suitably estimated. According to Vincent (1982), a typical storm spectra is often associated with a sharp peak frequency, f_m; consequently, a reasonable choice of f_c would be about 90 percent of f_m. The parameter α can be determined from the equation (5.13) together with a knowledge of f_m and the wind speed U. Further, when the frequency components containing the major part of the energy are in shallow water, as dictated by the condition $\omega_h < 1$, eq.(6.7) can be simplified to obtain

$$H_\ell = \frac{1}{\pi}(\alpha gh)^{\frac{1}{2}} f_c^{-1} \tag{6.8}$$

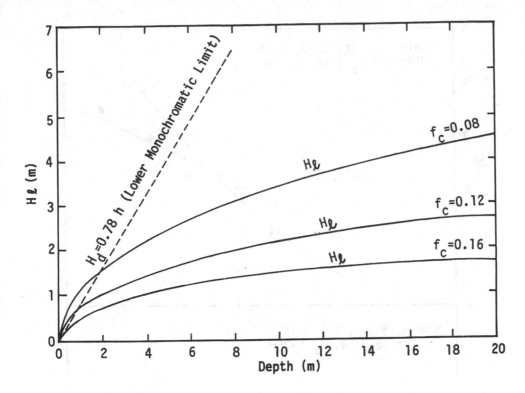

Figure 6.4: Depth-limited significant wave height, H_ℓ, as a function of water depth h and the cutoff frequency f_c. The curves are calculated for $\alpha = 0.0081$. H_d is plotted for the lower limit of 0.8h (from Vincent,1982).

Eq. (6.8) leads to an interesting consequence that H_ℓ varies with the square root of depth when the primary spectral components are depth-limited. Observational support for eq.(6.8) appears to be well provided by the data collected during the Atlantic Remote Sensing Land and Ocean Experiment (ARSLOE) conducted during October and November 1980. During this experiment, shallow water wave data were collected at Duck, North Carolina (U.S.A.) and were analyzed using eq.(6.8) as well as using a monochromatic theory which provides an estimate for the maximum wave height as $H_{max} \sim 0.5h$. As shown in Figure 6.5, eq.(6.8) provides a better fit to the shallow water wave data than does the monochromatic theory.

The similarity of wind wave spectrum in finite depth water has been examined recently by Bouws et al(1985) in the light of a paper by Kitaigorodskii et al(1975) on Phillips' theory of equilibrium range. The authors (Bouws et al) analyzed and combined three sets of shallow-water wave data collected in three separate studies; these were respectively: 1. Extreme storm wave data in the southern North Sea

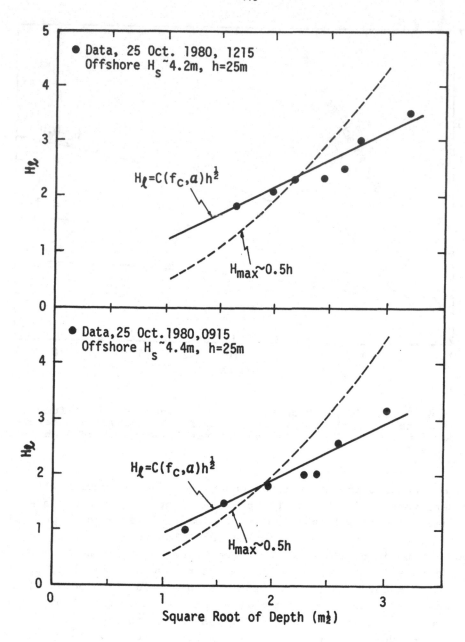

Figure 6.5: Variation of wave height with square root of depth for 25 October 1980 at Duck, North Carolina, U.S.A. The dots represent observed values, the solid line is based on measured α and f_C and the dashed line represented an estimated curve based on the monochromatic theory with H = 0.5 h (from Vincent, 1982)

117

collected near the lightship Texel, west of Rotterdam harbour, The
Netherlands, January 1976, 2. MARSEN (Marine Remote Sensing Experiment
at the North Sea) wave data in the German Bight, fall 1979 and 3.
ARSLOE (Atlantic Remote Sensing Land and Ocean Experiment) wave data
off North Carolina coast, U.S.A., fall 1980. Based on the analysis of
these data, Bouws et al hypothesized a shallow-water self similar
spectrum, E_{TMA}, given by

$$E_{TMA}(f,h) = E_J(f)\phi(\omega_h) \tag{6.9}$$

Here $E_J(f)$ is the JONSWAP spectrum and $\phi(\omega_h)$ is a ratio given by

$$\phi(\omega_h) = \frac{[(k(\omega,h))^{-3} \frac{\partial k}{\partial f}(\omega,h)]}{[(k(\omega,\infty))^{-3} \frac{\partial k}{\partial f}(\omega,\infty)]}$$

$$\text{and} \qquad \omega_h = 2\pi f(h/g)^{\frac{1}{2}} \tag{6.10}$$

Eq. (6.9) defines the well-known TMA (Texel, MARSEN, ARSLOE) spectrum,
a family of which is shown in Figure 6.6.

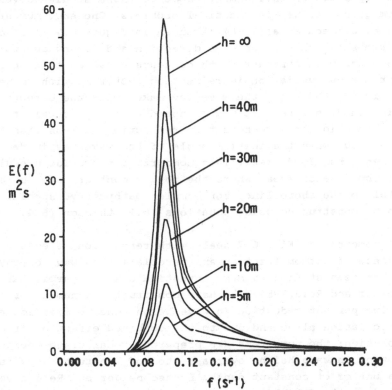

Figure 6.6: A family of TMA spectra with the same JONSWAP parameters
(f_m = 0.1 Hz, α = 0.01, γ = 3.3, σ_a = 0.07, σ_b = 0.09) but for
different water depths (from Bouws et al. 1985, Journal of Geophysical
Research, Vol. 90; Copyright by American Geophysical Union)

The curves of Figure 6.6 are drawn for different values of depth but with the same JONSWAP parameters as defined before (see eq. 5.15). The principal hypothesis used in defining the TMA spectrum is that the scaling factor of (6.9) is not restricted just to the saturation range but is valid across the entire spectrum.

The TMA spectrum, which has been shown to fit well a large number of measured shallow-water spectra can be interpreted as an upper limit to the finite depth spectra in wind seas propagating through sloping bottoms, typical of those over which the TMA spectrum has been defined. Further, the definition of the TMA spectrum has provided a useful technique to extend the application of deep-water spectral wave model to shallow-water regions.

6.4   Inclusion of Shallow-water Effects in Operational Wave Analysis and Prediction

A number of early studies summarized in section 6.2 were geared towards developing simplified techniques for calculating shallow water effects in operational environment. Some of these studies developed simple nomograms for use in practical problems. One such nomogram developed by Johnson et al(1948) is shown in Figure 6.7; this classical nomogram shows changes in wave direction and height due to refraction on beaches with straight depth contours parallel to a shore line. The nomogram shows variation in refraction factor K_r with respect to variation in θ^d, the angle the wave ray makes with the direction of propagation in deep water. It may be noted that on a straight shoreline, the reduction in wave height by refraction is less than 10 percent ($K_d = 0.90$) when the initial angle of the wave ray in deep water is less than 36 degrees. This nomogram can provide useful guidelines for those beach areas where the bottom contours are approximately parallel to the shore line. For complex bathymetry, special nomograms can be constructed using equations (6.1) through (6.4).

The nomogram of Fig. 6.7 deals with refraction effects only. The influence of bottom friction and percolation on wave energy dissipation has been studied by many investigators, in particular by Bretschneider and Reid(1954) who have constructed a number of nomograms giving percent reduction in wave height due to friction alone, due to percolation alone and due to the combined effect of friction and percolation; the nomograms are prepared to calculate the percentage reduction in wave height over a bottom of constant slope as well as over a bottom of constant depth. Bretschneider and Reid have also constructed a set of tables one of which was presented earlier (Table 6.I). These table values are calculated on the following assumptions: wave period is conserved; diffraction, reflection and supply of energy

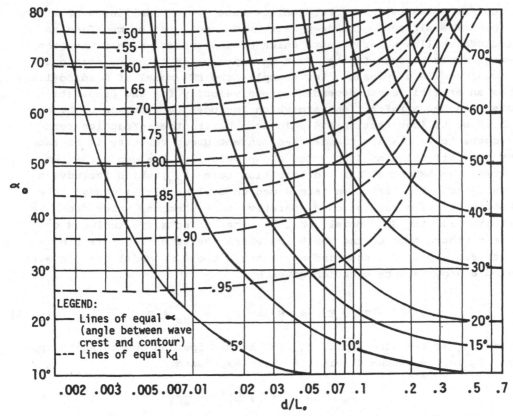

Figure 6.7: Changes in wave direction and height due to refraction on beaches with straight parallel depth contours. In this diagram, α_o refers to the angle the wave front makes with bottom contours in deep water, K_d refers to the refraction factor and α/L_o is the ratio of water depth (d) to the wavelength (L_o) in deep water. (from Johnson, O'Brian and Issacs, 1948).

wave period is conserved; diffraction, reflection and supply of energy are not considered; friction and permeability coefficients are assumed constant and steady-state conditions of transformations are assumed. With these limitations, the Table values can be interpreted as providing only an approximate solution to the problem of wave attenuation due to shallow-water effects. The Shore Protection Manual (1984) of the U.S. Army Coastal Engineering Research Centre is another source which provides a variety of calculations on shallow-water effects like refraction, diffraction, wave breaking etc. and also presents a number of nomograms showing shallow-water wave forecasting curves for various values of water depths.

The techniques discussed above are in general geared for applications to a site-specific location or over a small area like a beach

front or a coastal zone. For shallow-water wave modelling over a continental shelf zone, the deep-water wave model can be modified to
include shallow-water effects which become operative only when the
water depth becomes finite. Such an approach is taken by most operational models at present. For example, the BMO model of Golding(1983)
has an explicit depth-dependent group velocity for the propagation
step and has a refraction term which includes derivatives of h, the
water depth (see section 5.4). In deep water, the refraction term
reduces to zero while the depth-dependent group velocity is replaced
by the corresponding deep-water group velocity. Besides these two
terms, the BMO model has a dissipation term (S_{ds}) which includes a
quadratic bottom friction term based on Collins'(1972) study; this
bottom friction term becomes operative only when the water depth is
finite. For the GONO model, shallow-water effects are included over
the southern part of the North Sea where the water depth is finite
(see Fig. 5.8). For the wind-sea energy, the GONO model uses an averaged energy balance equation given by

$$\frac{\partial E}{\partial t} + \nabla \cdot (\bar{c}_g E) = S = S_{in} + S_{ds} + S_b \qquad (6.11)$$

Here S_{in} is the wind input, S_{ds} is the dissipation due to whitecapping
and S_b is the bottom dissipation term expressed as

$$S_b = -\Gamma \cdot \frac{\omega^2 \ E(\omega)}{g^2 \ \sinh^2 \ (\omega h)} \qquad (6.12)$$

In (6.12) Γ is the decay parameter whose value is determined based on
the JONSWAP data for swell. The combined effect of wind input and
whitecapping ($S_{in} + S_{ds}$) is determined by observations at infinite
fetch while an expression for average group velocity $\bar{c}_g$ in variable
water depth is developed. With this, eq. (6.11) can be solved over the
entire grid domain of the model; the shallow-water term S_b becomes
operative only when the water depth is finite.

The HYPA model described in section 5.3 has a shallow-water
version called HYPAS which uses the TMA spectrum defined by Bouws et
al(1985). The TMA spectrum changes over to the JONSWAP spectrum for
increasing water depth and the characteristic parameters of the TMA
spectrum are chosen to be the same as the JONSWAP parameters for deep
water. Further, the shape of the spectrum with a fixed peak frequency
is mainly determined by the balance of three source terms:

$$S_{in} + S_{n\ell} + S_{ds} = o \qquad (6.13)$$

Here S_{in} is the wind input term, $S_{n\ell}$ the nonlinear transfer term and
S_{ds} is the dissipation term due to whitecapping. It is assumed that
the dissipation mechanism S_{ds}, already present in deep water, varies

monotonically with wave number and this together with eg.(6.13) makes
the TMA spectral shape dependent upon the water depth. In deep water,
the energy balance as defined by eq.(6.13) leads to the JONSWAP
spectrum.

In the above examples, the shallow-water effects which are
modelled include refraction, shoaling, bottom friction and wave number
scaling as dictated by the TMA spectrum. The energy dissipation due to
wave breaking in shallow water is not directly included in any of the
above models. Two approaches to wave breaking in shallow water have
been put forward and modelled in recent studies (Resio, 1987; Young,
1988). The study by Young(1988) presents a second generation shallow-
water wave model in which five source functions are considered namely,
S_{in}, S_{nl}, S_{ds}, S_b and S_{brk}. Of these, the first three apply to deep-
water models and have already been discussed earlier. Young presents a
refined formulation of the S_{nl} term in which a scaling factor $A(\gamma)$, γ
being the peak enhancement factor, is developed based on JONSWAP data.
The terms S_{in}, S_{ds} and S_b are formulated by Young about the same way
as in other models, while the term S_{brk} is interpreted to represent
the depth-limited wave breaking for which the limiting wave height is
expressed as

$$\frac{H_s}{L} = 0.12 \tanh(kh) \tag{6.14}$$

Here H_s is the significant wave height, L is the wave length and h
is the water depth. Eq.(6.14) is reformulated in terms of the energy
spectrum to obtain a limiting value of the total energy E_{max} as

$$E_{max} = \frac{0.0355 \, \bar{c}^4}{g^2} \tag{6.15}$$

Here $\bar{c}$ is the phase speed of the mean frequency of the spectrum.
Eq.(6.15) places a limit on the total energy of the spectrum but gives
no information of the limiting spectrum itself. As long waves 'feel'
the bottom to a greater extent than the short ones, depth-limited wave
breaking would be expected to occur first at low frequencies moving
progressively to higher frequencies with decreasing depth. Eq.(6.15)
was used by Young to adjust the energy levels of various frequencies
in his model, which was applied on a nested grid to simulate wave
conditions generated during the passage of selected tropical cyclones
in the Indian Ocean off the west coast of Australia. The model appears
to perform quite well even under demanding meteorological conditions.

The shallow-water wave model of Resio(1987) is an extension of
his earlier (Resio,1981) deep-water wave model and is based on the
hypothesis that in shallow waters of the nearshore regions, wave

breaking rather than bottom friction is the primary mechanism for wave height decay. This hypothesis appears to be substantiated by recent studies, for example, by Thornton and Guza(1983) who estimate that only about 3% of the total variation in wave height is associated with bottom friction. Resio further postulates that away from the shoreline, wave breaking is primarily 'steepness-limited' while in the very shallow water near the shoreline, 'depth-limited' breaking becomes the dominant wave-breaking mechanism. Starting from the fundamental form of the collision integral for four reasonantly interacting waves (Hasselmann,1962), Resio develops a theoretical framework for the characteristic form of equilibrium spectra in waters of arbitrary depth. An important consequence of this spectral equilibrium formulation is that a strong constant flux of wave energy exists through the equilibrium range towards the high frequency region where it is lost due to wave breaking. A dynamic balance between wind input and non-linear flux dominates the shape of the spectrum and controls energy level and energy loss in waves propagating to water depths of 10 m or less. The bottom friction is considered as a free parameter in the shallow-water model of Resio and is included only for long period swell waves with little or no wind. In the follow-up study, Resio (1988) applied his model to data sets from a wide range of physical scales (laboratory through oceanic) as well as from several locations around the world and obtained spectral shapes which agreed well with those measured under storm conditions. Based on these studies, Resio has developed an operational model called WAVAD which uses a coarse grid (2° latitude x 2° longitude grid spacing) covering the western north Atlantic ocean and two nested grids covering the Scotian shelf region in the Canadian Atlantic. Of the two nested grids, the intermediate grid has a spacing of 0.25° in latitude and longitude while the fine grid has a spacing of 4 nautical miles (see Figure 6.8). The WAVAD model was used in a wave model intercomparison study during the CASP field project January-March 1986.

The operational ODGP model has a shallow-water algorithm developed by Oceanweather Inc. This algorithm includes the shallow-water processes of refraction, shoaling, bottom friction and wave-number scaling. Refraction and shoaling are modelled within the propagation step, while bottom friction and wave number scaling are modelled within the growth step. The principal modifications to the deep-water wave growth algorithm are a. transformation of the fully developed P-M spectrum form to shallow water, b. calculation of an explicit attenuation associated with bottom friction which is modelled after a comprehensive treatment of Grant and Madsen(1982), c. calculation of the exponential wave growth rate using the shallow-water celerity and d. employing wave number scaling of the high-frequency saturation range of the spectrum, with the equilibrium range coefficient α, taken as a

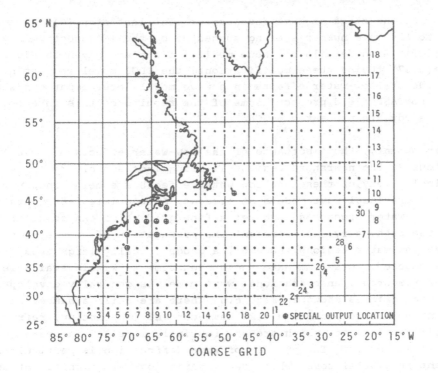

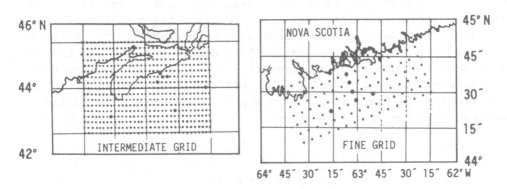

Figure 6.8: <u>Top</u>: The coarse grid (2° x 2°) for the WAVAD model. <u>Bottom</u>: The intermediate (0.25° x 0.25°) and the fine (4 nm everywhere) grid for the WAVAD model (source: Dr. D. Resio)

function of the stage of wave development. The last step (d) is equivalent to the wave number scaling suggested by the TMA spectrum given by eq.(6.9). More details of these modifications are given in Eid and Cardone(1987). Using these modifications, the ODGP model was extended to include shallow-water effects in a wave model intercomparison study during the CASP field project; some of the results of this intercomparison study are presented in Chapter 7.

In summary, the importance of shallow-water effects on wave conditions in the nearshore regions is being increasingly realized; accordingly, many of these shallow-water effects are being incorporated in operational wave models at present. Appropriate representation of shallow-water processes requires a fine grid which can adequately define the bathymetry. To represent the propagation of wave energy into shallow-water regions, however, a geographically large area must be modeled. To accommodate these requirements within available computer resources, an operational wave model is generally developed on a nested grid as shown in Fig. 6.8. There are still a number of important nearshore processes like wave-current interaction, refraction, diffraction etc. which are not routinely modelled in operational wave analysis so far. The problem of wave diffraction is particularly important in coastal zones where depth variations are significant and where strong currents are known to exist. In such a situation, the combined effect of refraction and diffraction has been modelled using the mild slope equation first developed by Berkhoff(1972). Many recent studies (ex. Kirby and Dalrymple, 1983; Kirby et al., 1984; Liu, 1982) have used the mild slope equation or its modification to investigate the combined refraction-diffraction effects in shallow waters. These and other nearshore processes are expected to receive increasing attention in the near future.

REFERENCES

Aranuvachapun, S. 1977: Wave refraction in the southern North Sea. Ocean Engineering, 4, 91-99

Berkhoff, J.C.W., 1972; Computation of combined refraction-diffraction. Proc. 13th Coastal Engineering Conference, Vancouver 1972, ASCE, New York, Vol. 1, Ch. 24, 471-490.

Bouws, E., H. Gunther, W. Rosenthal and C.L. Vincent, 1985: Similarity of the wind wave spectrum in finite depth water, Part 1 - spectral form. J. Geophysical Research, 90, C1, 975-986.

Bretschneider, C.L. and R.O. Reid, 1954: Changes in wave height due to friction, percolation and refraction. Tech. Mem. Beach Erosion Board, U.S. Army Corps of Engineers, No. 45, 36 pp.

Cavaleri, L. and P.M. Rizzoli, 1981: Wind wave prediction in shallow water: Theory and applications. J. Geophysical Research, 86, 10961-10973

Collins, J.I. 1972: Prediction of Shallow-water spectra. J. Geophysical Research, 77, 2693-2707.

Eid, B.M. and V.J. Cardone, 1987: Operational test of wave forecasting models during the Canadian Atlantic Storms Program (CASP). Environmental Studies Research Funds, Report No. 076, Ottawa, 111 p+ Appendices.

Golding, B. 1983: A wave prediction system for real-time sea state forecast. Q. J. Royal Met. Society, 109, 393-416.

Grant, W.D. and O.S. Madsen, 1982. Movable bed roughness in unsteady oscillatory flow. J. Geophysical Research, 87, C1, 469-481.

Griswold G.M., 1962: Numerical calculation of wave refraction. J. Geoph. Research, 68, 1715-1723.

Hays, J.G., 1977: Ocean current and shallow water wave refraction in an operational spectral ocean wave model. Ph. D. Thesis, New York Univ. 166 pp.

Hays, J.G., 1980: Ocean current wave interaction study. J. Geophysical Research, 85, C9, 5025-5031.

Hasselmann, K., 1962: Loc. cit. (Ch. 3)

Johnson, J.W., 1947: The refraction of surface waves by currents. Trans. Amer. Geoph. Union, 28, 867-874.

Johnson, J.W., M.P. O'Brien and J.D. Isaacs, 1948: Graphical construction of wave refraction diagrams. H.O. Pub. No. 605, U.S. Navy Hydrographic Office, Washington, D.C. January 1948, 45 pp.

Jonsson, I.G. and J.D. Wang, 1980: Current-depth refraction of water waves. Ocean Engineering, 7, 153-171.

Karlsson, T., 1969: Refraction of continuous ocean wave spectra. J. Waterways, Harbours, Coastal Engineering Division, 95(WW4), 437-448.

Kirby, J.T. and R.A. Dalrymple, 1983: A parabolic equation for the combined refraction-diffraction of Stokes waves by mildly varying topography. J. Fluid Mechanics, 136, 453-466.

Kirby, J.T., P.L.-F. Liu, S.B. Yoon and R.A. Dalrymple, 1984: Combined refraction-diffraction of nonlinear waves in shallow water. Proc. 19th Coastal Engineering Conference, Houston, 1984, ASCE, New York, Vol. 1, Ch. 68, 999-1015.

Kitaigorodskii, S.A. et al. 1975: Loc. cit. (Ch. 5)

Liu, P.L.-F., 1982: Combined refraction and diffraction: comparison between theory and experiments. J. Geophysical Research, 87, 5723-5730.

Lambrakos, K.F., 1981: Wave-current interaction effects on water velocity and surface wave spectra. J. Geophysical Research, 86, C11, 10955-10960.

Longuet-Higgins, M.S., 1956: The refraction of sea waves in shallow water. J. Fluid Mechanics, 1, 163-176.

Munk, W.H., and R.S. Arthur, 1952: Wave intensity along a refracted ray. Natl. Bureau of Standard Circular, U.S.A., 521, 95-108.

Putnam, J.A. and J.W. Johnson, 1949: The dissipation of wave energy by bottom friction. Trans. Amer. Geophysical Union, 30, 349-356.

Resio, D.T., 1981: Loc. cit. (Ch. 5)

Resio, D.T., 1982; Wave prediction in shallow water. Presented at the 14th Annual Offshore Technology Conference, Houston, U.S.A., May 1982, Paper No. 4242.

Resio, D.T., 1987: Loc. cit. (Ch. 5)

Resio, D.T., 1988: Shallow-water waves, Part II: Data Comparisons. J. Waterway, Port, Coastal and Ocean Engineering, 114, 1, 1-16.

Sanderson, R.M., 1974: The unusual waves off southeast Africa. Marine Observer, 44, 180-183.

Shemdin, O.H., et al, 1980: Mechanisms of wave transformation in finite-depth water. J. Geophysical Research, 85, C9, 5012-5018.

Shiau, J.C. and H. Wang, 1977: Wave energy transformation over irregular bottom. J. Waterway, Port, Coastal and Ocean Engineering, 103 (WW1), 57-63.

Shore Protection Manual, 1984: U.S. Army Coastal Engineering Research Centre, Vicksburg, U.S. Govt. Printing Office, Washington, D.C. Volume 1 (fourth Edition).

Sturm, H., 1974: Giant waves. Ocean, 2(3), 98-101.

Thornton, E.B. and R.T. Guza, 1983: Transformation of wave height distribution. J. Geophysical Research, 88, C10, 5925-5938.

Vincent, C.L., 1979: The interaction of wind-generated sea waves with tidal currents. J. Physical Oceanography, 9, 748-755.

Vincent, C.L., 1982: Depth-limited significant wave height: A spectral approach. Tech. Rep. No. 82-3, U.S. Army Corps of Engineers, Coastal Eng. Res. Centre, August 1982, 23 pp.

Wang, H. and W.-C. Yang, 1981: Wave spectral transformation measurements at Sylt, Nort Sea. Coastal Engineering, 5, 1-34.

Wilson, W.S., 1966: A method for calculating and plotting surface wave rays. Tech. Memo. No. 17, U.S. Army Coastal Eng. Res. Centre.

Young, I.R., 1988: A shallow water spectral wave model. J. Geophysical Research, 93, C5, 5113-5129.

CHAPTER 7
VALIDATION OF WAVE MODELS

7.1 General Comments

With increasing number of ocean wave models having been developed in recent years, appropriate evaluation of a wave model against
observed wind and wave data becomes necessary as well as important. A
spectral wave model can be tested in two ways. First, the model can be
used to simulate the evolution of wave spectrum with fetch or duration
for stationary, uniform wind fields. A more revealing test would be to
determine the model's ability to specify the directional wave spectrum
for realistic wind fields that vary in space and time. This is a
difficult test to carry out for at least three reasons: $\underline{1}$. it is
difficult to specify accurate wind fields over the oceans; $\underline{2}$. very few
measurements of directional wave spectrum are available at present,
and $\underline{3}$. where wave measurements are available, the spectral analysis
provides only an estimate of the true spectrum. Besides these problems, there are uncertainties due to sampling variability of the
measured spectra. consequently, most evaluation studies on wave models
have considered significant wave height (H_s), significant (or peak)
wave period (T_s) and wind (speed and direction) as important parameters for evaluation against observed data. A number of evaluation
studies on individual wave models or group of models have been reported by various investigators in recent literatures. Some of these
studies are summarized below.

7.2 Evaluation of Individual Wave Models

An important requirement for appropriate evaluation of a wave
model is the availability of reliable sea-state measurements and
related weather data. The severe storm of December 1959 in the north
Atlantic generated high waves which were recorded by the ship-borne
wave recorder of the weathership 'Weather-Reporter' at station
J(52.5°N, 20°W). The weathership also provided adequate weather data
for the period 15-21 December 1959 which covered the storm period. The
availability of good weather and wave data prompted Feldhausen et al
(1973) to evaluate wave hindcasting procedures of Bretschneider(1963),
Barnett(1968), Inoue(1967), Pierson, Tick and Baer(1966) and Wilson
(1955); these wave hindcasts were compared against the recorded significant wave heights as well as against visually observed wave heights

as shown in Figure 7.1a. The recorded significant wave heights in this Fig. are displayed in terms of their 90% confidence limits, while the shaded area shows the range of wave hindcast values calculated using various procedures; also shown in the Fig. is the variation of visually observed wave heights during the storm period. In general, the calculated wave heights are within the 90% confidence limits of the recorded wave heights. A linear regression between the measured wave heights and the calculated wave heights was obtained for each of the procedures as mentioned above. In Figure 7.1b are shown two regression lines together with data points generated using the models of Barnett and Bretschneider respectively. These two regression lines provide the best fit between measured and calculated wave heights with correlation coefficient values of 0.87 and 0.92 respectively.

Of the wave hindcasting procedures used in the above procedure, two were based on the 'significant wave' method, while the remaining three employed the spectral character of the sea-state to generate hindcasts of significant wave height. In the following example, we present the evaluation of hindcast and measured one-dimensional spectrum at a rig location in the Gulf of Mexico during the passage of hurricane Camille on 17 August 1969. The hindcast spectrum was based on a version of the PTB spectral wave model which was calibrated against several hurricane wind fields during the Ocean Data Gathering Program (Cardone et al. 1976); this model has been upgraded and is now identified as the ODGP model. Figure 7.2a shows the hindcast and the measured one-dimensional spectrum together with the 90% confidence limits of the measured data. The measured and hindcast spectra are in close agreement with each other providing significant wave height values within two feet of each other. Another evaluation of the same model based on observations of maximum wave heights associated with selected hurricanes in the Gulf of Mexico is shown in Figure 7.2b. The wave heights ranged from 6 to 24 m (20 to 79 ft) and the RMS difference between measured and hindcast wave heights was 1.5 m(4.9 ft). Additional evaluation results for the operational ODGP model are discussed in a later section.

Pierson(1982) has presented several examples of verification of the U.S. Navy's SOWM (Spectral Ocean Wave Model). Figure 7.3a shows an example of SOWM verification at a grid point in the north Atlantic during the severe weather conditions from 25 November to 14 December 1966. Here the wind speed and the significant wave height generated at the SOWM grid point 128 were compared with recorded wind speed and wave heights at Station India (59°N, 19°W) located about 53 km from the grid point 128. Figure 7.3b shows a scatter diagram for 211 pairs of significant wave heights at Station India and at the grid point (wave height values at station India were based on a Tucker wave gauge

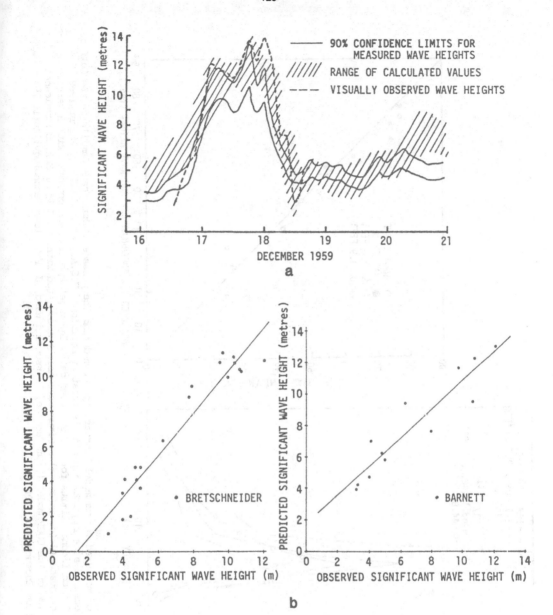

Figure 7.1: a. A comparison of measured versus calculated significant wave heights obtained using different models. b. Scatter diagrams and regression lines between observed and predicted significant wave heights using models of Bretschneider and Barnett (from Feldhausen, Chakrabarti and Wilson, 1973)

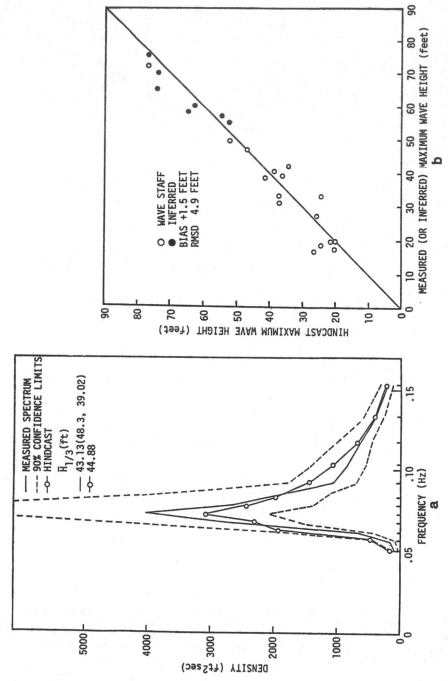

Figure 7.2: a. Comparison of measured versus model hindcast spectrum during the passage of hurricane Camille in the Gulf of Mexico. The spectrum was measured at a rig location off the coast of Louisiana, U.S.A., at 1600 CDT, 17 August 1969. The dashed lines indicate the 90% confidence limits for the measured data. The model spectrum was obtained with a version of the ODGP model. b. Comparison of measured (or inferred) maximum wave height associated with the Gulf of Mexico hurricanes and hindcasts made with a version of the ODGP model (from Cardone, Pierson and Ward, J. Petroleum Technology, April 1976; Copyright by Society of Petroleum Engineers of AIME, U.S.A.)

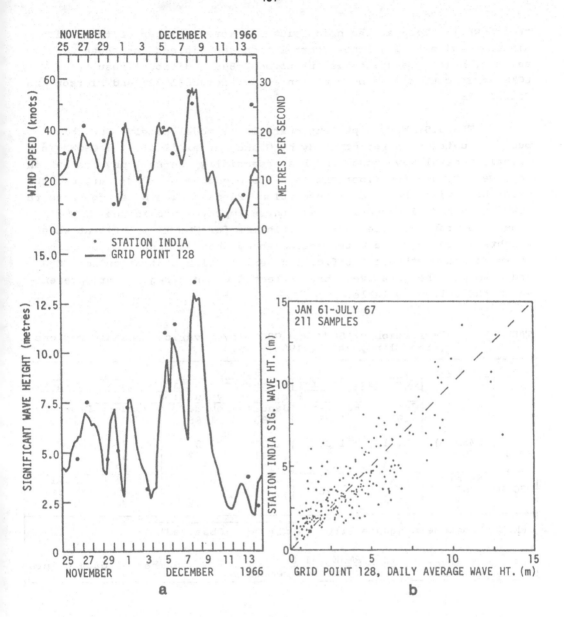

Figure 7.3: a. A comparison of wind and wave data between station India and the SOWM grid point 128 during severe weather conditions from 25 November to 14 December 1966. b. Comparison of daily average significant wave height between station India and SOWM grid point 128. See Fig. 5.1 for the location of grid point 128 (from Bales, Cummins and Comstock, 1982; Copyright the Society of Naval Architects and Marine Engineers: Reprinted with permission).

record while those at the grid point 128 were averages of four daily hindcast values). The linear correlation coefficient for these 211 pairs of height values comes out to be 0.84; additional results of SOWM verification have been presented by Pierson(1982) and Pierson and Salfi(1979).

The U.S. Navy's present operational model namely GSOWM has been evaluated in a recent study by Clancy et al(1986). The GSOWM is a global spectral wave model and has essentially the same physics but an improved propagation algorithm than its predecessor SOWM; further, the GSOWM has twice the angular resolution (24 direction bins) compared to SOWM which had 12 direction bins; these improvements in the GSOWM appear to provide better error statistics for GSOWM significant wave heights which were evaluated against buoy observations in the north Atlantic and northeast Pacific. The verification statistics for GSOWM and SOWM wave heights over three selected oceanic regions are present- ed in the following Table.

TABLE 7.I.   Evaluation of SOWM and GSOWM over various oceanic regions (from Clancy et al. 1986)

Model	Parameter	Regions					
		Northeast Pacific		Northwest Alantic		Northeast Atlantic	
		Period of Evaluation					
		January 1985	February 1985	January 1985	February 1985	January 1985	February 1985
SOWM	RMSE(m)	1.71	1.35	1.31	1.21	2.76	2.61
	SI	47	38	43	52	59	53
	N	338	386	190	102	98	94
GSOWM	RMSE(m)	1.27	0.88	0.95	0.73	1.86	1.71
	SI	34	28	29	32	40	34
	N	338	386	190	102	98	94

RMSE = Root Mean Square Error = $\sqrt{\frac{1}{N}\Sigma(\text{Model}-\text{Observed})^2}$

SI   = Scatter Index = $\dfrac{\text{RMSE} \times 100}{\text{Mean observed value}}$; N = number of data points

It can be seen from the above Table that in general the RMS errors in the GSOWM significant wave heights are reduced by about 30 percent when compared against the corresponding error for SOWM wave heights. Both the models produce relatively large errors over the northeast Atlantic where the wave observing buoys are located over regions which often experience large wind-generated waves as well as swells during the winter months.

The GONO model has been tested at four waverider locations (see Fig. 5.6) for several storm cases in the North Sea (Janssen,

Komen and De Voogt, 1984). A sample verification at two locations
(Pennzoil and Euro) is shown in Figure 7.4 for the storm period 19-22
April 1980; this storm provided extreme long fetches for the two waverider locations. The variation in significant wave height for the
duration of the storm period is shown in Figure 7.4a; overall, the
model generated wave heights agree quite well with the measured
values, although the model generated wave height maxima appear to lag
the observed maxima by about six hours or so. In Figure 7.4b are shown
observed and computed spectra for three selected hours of 20 April
1980. At both the locations, strong dissipation terms (due to bottom
effects) appear to produce significant underestimates of the spectral
values at low (swell) frequencies during the first 12 hours; this
underestimation of model generated energy is due to the fact that the
bottom dissipation term dominates the wind input term at both the
locations during the peak of the storm. In another study (Günther,
Komen and Rosenthal, 1984), the GONO model was evaluated along with
the HYPAS (a shallow-water version of HYPA) model at three locations
in the North Sea; these locations are shown in the GONO model grid of
Fig. 5.6. Both the GONO and the HYPAS are based on very similar concepts namely a parametrical wind/wave representation and a characteristic ray method for swell propagation. Both the models were tested on
the GONO grid with a grid spacing of 75 km everywhere and both were
driven by the same wind fields namely the operational wind field of
GONO. The error statistics for the two models during the test period
December 1979 to April 1980 are shown in the following Table.

TABLE 7.II: **Evaluation of HYPAS and GONO at three locations in the
North Sea (from Günther et al, 1984)**

Location	Parameter	December 1979		January 1980		April 1980	
		HYPAS	GONO	HYPAS	GONO	HYPAS	GONO
Euro	ME	0.05 m	-0.04	0.17	-0.11	0.10	-0.06
	RMSE	0.53	0.55	0.43	0.51	0.34	0.40
	SI	26	27	34	40	28	33
Ijmuiden	ME	0.01	-0.01	0.26	0.02	0.11	-0.05
	RMSE	0.52	0.53	0.40	0.52	0.39	0.46
	SI	24	25	25	33	28	34
Pennzoil	ME	-0.06	-0.03	0.03	-0.11	0.09	0.02
	RMSE	0.51	0.55	0.43	0.55	0.44	0.49
	SI	22	24	27	34	29	33

ME : Mean Error $= \frac{1}{N}\Sigma$ (Model-observed)

        The error statistics of Table 7.II suggest that both the GONO
and the HYPAS models appear to perform equally well at the three North
Sea locations; further, the error statistics for GONO and HYPAS (in
Table 7.II) are in general smaller than the error statistics for SOWM

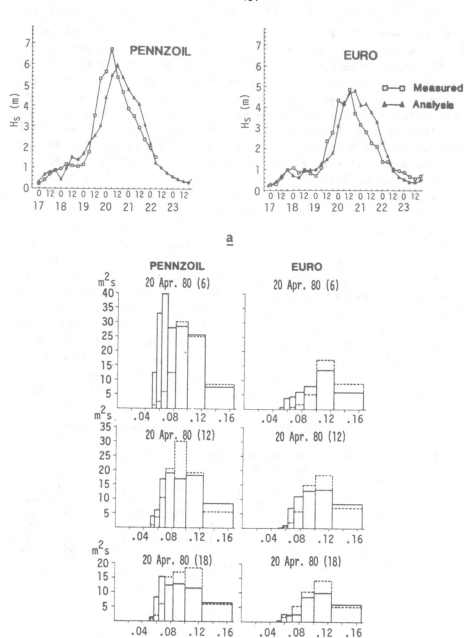

Figure 7.4: Evaluation of the GONO model for the storm period 19-22
April, 1980. a. Variation in significant wave height at two locations
PENNZOIL and EURO. b. Observed versus computed spectra for 20 April
1980 at the same two locations (from Janssen, Komen and De Voogt,
Journal of Geophysical Research. Vol. 89; Copyright by American
Geophysical Union)

and GSOWM (in Table 7.I). This reduction can be attributed to at least
two factors: 1. The models SOWM and GSOWM are based on first generation physics whereas the models GONO and HYPAS are based on second
generation physics; furthermore, both GONO and HYPAS have shallow
water effects explicitly included in them; 2. The grid spacing for
SOWM and GSOWM is at least 200 km or more, whereas the GONO and HYPAS
operate on a 75-km grid spacing. Besides these factors, improved wind
specification may also contribute to reduction in error statistics for
GONO and HYPAS.

The operational global model of the BMO (British Meteorological Office) is routinely verified against the wind and wave data
from buoys, oil platforms and ocean weather ships which are transmitted on the Global Telecommunication System in near-real time. About 35
stations, all in the northern hemisphere, are used in this routine
verification. The verification is done at every 6 hours for analysis
fields and at every 12 hours for forecast fields. Table 7.III presents
the BMO model verification for the month of January 1989.

Table 7.III: **Verification statistics for the global BMO model for
January 1989**

fore-cast time	Para-meter	WIND SPEED (ms^{-1}) Range of values (m s^{-1})					WAVE HEIGHT (m) Range of values (m)				
		0-10	10-15	15-20	>20	All	0-3	3-6	6-9	>9	All
00 hr	ME	0.2	0.2	-0.3	-1.2	0.1	-0.1	-0.7	-0.8	-0.4	-0.3
	RMSE	1.9	2.4	2.4	3.6	2.0	0.7	1.2	1.8	2.0	1.0
	N	2381	654	162	57	3254	2057	1113	201	36	3407
12 hr	ME	0.9	1.3	-0.5	-2.3	0.9	0	-0.4	-0.5	-0.7	-0.2
	RMSE	2.9	3.6	3.8	5.7	3.1	0.8	1.2	2.0	2.0	1.0
	N	1177	334	77	31	1619	1021	558	100	16	1695
24 hr	ME	1.2	1.2	-0.3	-3.0	1.0	0.1	-0.2	-0.4	0	-0.1
	RMSE	3.1	3.9	4.5	6.0	3.5	0.8	1.2	2.0	2.3	1.1
	N	1177	334	77	31	1619	1021	558	100	16	1695
36 hr	ME	1.3	0.8	-1.0	-3.1	1.0	0.1	-0.2	-0.5	0.4	0
	RMSE	3.4	3.9	4.7	6.8	3.7	0.8	1.3	2.0	2.4	1.1
	N	1177	334	77	31	1619	1021	558	100	16	1695

It can be seen from the above Table that in general, the RMS
error increases steadily as the range of values for wind speed and
wave height increases. This increase (decrease) in the RMS error with
increasing (decreasing) values of wind speed and wave height is also
reflected in the performance statistics of the BMO model in other
months (or seasons). For example, for the period July-September 1988,
the RMS error for wind speed was found to be 1.6 m s^{-1} in the Pacific

and about 1.8 m s^{-1} in the Atlantic, while for significant wave height, the RMS error was 0.6 m in the Pacific and about 0.4 m in the Atlantic. Overall, the BMO model generates error statistics which show an improvement of about 25 percent over the corresponding error statistics for the GSOWM presented in Table 7.I.

The third generation WAM model which is being executed daily at the ECMWF in Reading, United Kingdom since spring 1987 has been evaluated against wind and wave data from moored buoys in a recent study (Zambresky, 1989). The WAM model evaluation has been made for a one-year period from December 1987 to November 1988 and error statistics over four areas namely the Hawaiian Islands, Gulf of Alaska, east coast of U.S.A. and northeast Altantic (north of the United Kingdom) have been prepared as shown in Table 7.IV. The evaluation of wind and wave products is made at analysis time (zero hour forecast) only, since the WAM model generates only 24-hour forecasts at this time.

Table 7.IV: **Verification statistics for the WAM model for the period December 1987-November 1988; zero-hour forecast (from Zambresky, 1989)**

Parameter	REGION			
	Hawaiian Islands	Gulf of Alaska	East Coast of U.S.A.	Northeast Atlantic
WIND SPEED (m s^{-1})				
ME	-0.4	0.5	0.1	-0.5
RMSE	1.2	2.0	1.9	2.4
SI	18	27	30	28
r	0.82	0.87	0.83	0.81
N	2057	4551	4229	2455
WAVE HEIGHT (m)				
ME	-0.28	-0.22	-0.38	-0.40
RMSE	0.47	0.67	0.66	0.82
SI	22	22	37	29
r	0.78	0.91	0.83	0.85
N	2061	4657	4284	2825
PEAK PERIOD (s)				
ME	-0.01	-0.70	-0.71	data not available
RMSE	1.99	1.92	1.44	
SI	21	18	19	
r	0.54	0.67	0.66	
N	648	505	903	
r : linear correlation between model and observed value				

The WAM model is being driven by winds at 10 m level generated by the global operational weather prediction model of the ECMWF. As can be seen from Table 7.IV, the ECMWF winds compare quite favourably against the moored buoy data with a bias varying from +0.5 m s^{-1} to -0.5 m s^{-1} and a RMS error of less than 2 m s^{-1} on the whole. The wave height error statistics produced by the coarse grid (3° x 3°) WAM model are comparable to those in Table 7.II which are generated by the finer grid models like HYPAS and GONO. An improved wind specification together with inclusion of the third-generation source terms may be attributed to the improved error statistics obtained for the WAM model.

7.3 Intercomparison of Wave Models

An intercomparison of two or more wave models has been carried out by several scientists working either individually or as a team. Two approaches to intercomparison of wave models have been taken so far. In one approach, a number of wave models are tested using hypothetical wind fields representative of typical atmospheric flow patterns. In the second approach, wave models are driven using real wind fields at selected time periods and the model results are evaluated against available wind and wave observations. Both approaches have certain advantages and disadvantages some of which will be considered in the following discussion.

7.3.1 Intercomparison with Simulated Wind Fields

One of the earliest studies on intercomparison of wave models has been reported by Dexter(1974) who used three models, 1. Wilson's significant wave model; 2, a version of the PTB spectral model and 3, Barnett's spectral model. Dexter integrated the three models using a test wind speed of 15 ms^{-1} which was kept constant during the first 30-hour integration period; the wind was then switched off for the next 30 hours so that the integration could simulate wave growth as well as wave decay in the models. The time histories of significant wave heights for the three models are shown in Figure 7.5. For comparison, the significant wave height growth based on the classical SMB model is also shown in the Figure. Several interesting observations can be made from Fig. 7.5; a. The wave heights for the Barnett model are, in general, highest throughout the wave growth period; b. The wave growth for the PTB model is the fastest and attains a 'fully developed' stage in about 18 hours; c. The decay for the Barnett and the PTB models are similar but tend to diverge; this is due to the nonlinear terms in the Barnett model which allow continued wave growth at low frequencies and this results in slower decay; d. In general, there is a good agreement between Wilson's significant wave model and the two spectral wave models.

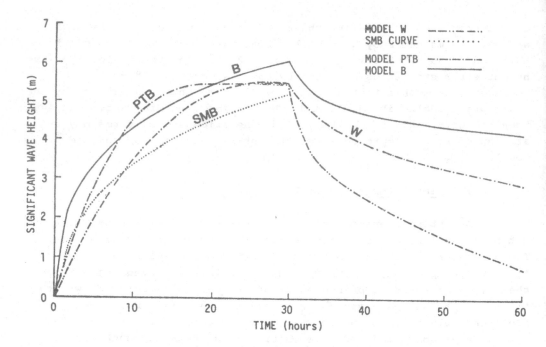

Figure 7.5: Time histories of significant wave heights for three models, Wilson's model (W), PTB model and Barnett's model (B). A test wind speed of 15 m s^{-1} is used from 0 to 30 hr after which the wind is switched off for the remaining duration. For comparison, the growth curve obtained by the SMB nomogram is shown by the dotted line (from Dexter, 1974; Copyright by American Meteorological Society)

Dexter's study, although interesting, does not provide any insight into the operational utility of these models.

The most comprehensive wave model intercomparison study using simulated wind fields has been reported by the SWAMP Group (1985). In this intercomparison study, ten spectral wave models were tested using several idealized wind fields representative of typical atmospheric flow patterns like uniform stationary wind blowing orthogonally off a straight coastline, uniform winds blowing at an angle to the coastline, a wind field with sudden change in wind direction representative of a frontal passage and wind fields representative of stationary and moving hurricanes. The ten spectral models were classified into three catagories namely DP (Decoupled Propagation), CH (Coupled Hybrid) and CD (Coupled Discrete) models. All the ten models were run on identical grids as far as possible and in all test cases (excepting that of a moving and a stationary hurricane) a nominal reference wind speed U_{10} (at 10 m height above water level) of 20 m s^{-1} was assumed. The model variables were all transformed to nondimensional form by the following expressions:

Nondimensional fetch $\qquad x^* = gx/U_*^2$

Nondimensional time $\qquad t^* = gt/U_*$ $\qquad\qquad$ (7.1)

Nondimensional peak frequency $\qquad f_p^* = U_* f_p g^{-1}$

Nondimensional energy $\qquad E^* = g^2 E U_*^{-4}$

In the above, U_* is the friction velocity which is defined
earlier as $U_* = U\sqrt{C_D}$; here C_D is the drag coefficient for air, which
in general depends upon the local atmospheric boundary layer and the
sea-state parameters. The SWAMP study assumed a drag coefficient of
1.83×10^{-3} for all models. The various energy source functions (S_{in},
S_{nl}, S_{ds}) were calculated using expressions prescribed for each of
the ten models. The nonlinear transfer function S_{nl} was calculated in
respect of four CD models; of these four models, the EXACT-NL model
used the exact computations of S_{nl} for a simplified wave field geometry.

The SWAMP study obtained a large number of results using simulated wind fields as mentioned earlier. Of particular interest to our
discussion on wave model intercomparison are the results of the fetch
and duration-limited growth curves for various models; these growth
curves were obtained by specifying for each model a stationary, homogeneous wind field with wind speed $U_{10} = 20$ m s^{-1} blowing orthogonally
offshore and numerically integrating each model till a stationary
state was reached. For ready reference, the growth curves produced by
limited growth curves showing the variation of energy (E^*) and peak
frequency (f_p^*) with distance (x^*) are shown in Fig. 7.6a and 7.6b,
while the corresponding duration limited growth curves are shown in
Fig. 7.7a and 7.7b respectively. Also shown in these Figures are
growth curves for two other models namely SOWM and ODGP; these two
growth curves were obtained in an evaluation study initiated by
Environment Canada (see MacLaren Plansearch Ltd. 1985). The growth
curves for the SWAMP models appear to exhibit strong differences among
different models inspite of the fact that all the models were calibrated against fetch or duration limited data. These differences are
attributed to the indeterminate nature of the drag law relating U_* to
U_{10} and to the uncertainty as to whether U_* or U_{10} is the more appropriate variable for describing wave growth. Further, it is speculated
that the gustiness of the wind or the background turbulence may contribute to the observed difference in wave growth rates. Considering
further the growth curves for SOWM and ODGP, it appears that at the
shortest fetches both the SOWM and the ODGP models exhibit more energy
(Figure 7.6) and lower peak frequencies than the SWAMP models; this
behaviour may be attributed to the relatively low resolution of the
higher frequencies in both the SOWM and ODGP models compared to the

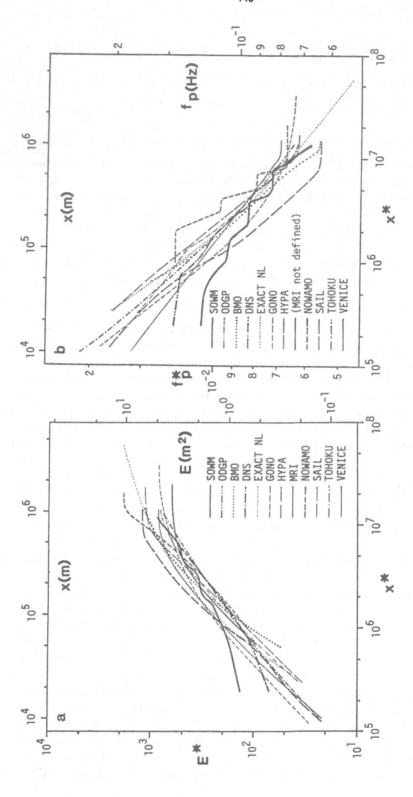

Figure 7.6: a. Nondimensional fetch-limited growth curves for the total energy E* and b. Nondimensional fetch-limited growth curves for the peak frequency f_p^* in respect of 12 spectral wave models (from The SWAMP Group, 1985 and MacLaren Plansearch Ltd., 1985)

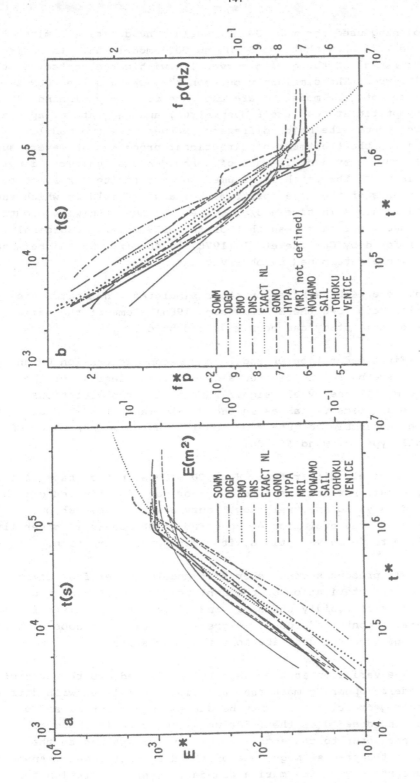

Figure 7.7: Nondimensional duration-limited growth curves for a. total energy E* and for b. peak frequency f_p^* in respect of same 12 models as in Fig. 7.6. (from The SWAMP Group, 1985 and MacLaren Plansearch Ltd., 1985)

frequency binning used for most SWAMP models. The duration limited tests (Figure 7.7) show that the SOWM and ODGP models have the same basic growth rates and the growth curves lie within the scatter of the SWAMP model curves. The similarity between the SOWM and the ODGP is expected since both the models share the same frequency binning, A and B terms, asymptotic states (the P-M spectrum) and the rate of approach to full development. The major difference between the two models' algorithms is in the treatment of directional processes; however, such effects will not be evident in case of a homogeneous, unidirectional wind field in which the fetch is assumed to be infinite for all directional components of the wave spectrum. For a wind field in which the wind direction varies in time and/or space, the duration-wise growth of the SOWM becomes faster than that of the ODGP model; accordingly, the SOWM was found by Cardone et al (1976) to specify much higher sea states than those determined by observations.

Many more results based on other simulated wind fields are discussed elsewhere (see the SWAMP Group, 1985). Some of the main conclusions of this study can be summarized below:

1. First generation DP models and second-generation CH and CD models, which are based on fundamentally different designs of the spectral energy balance, yield significantly different relations between space and time variables in the development of a wind-sea. It is therefore impossible to tune a DP model to agree with a CH or CD model for all types of wind fields.

2. DP models differ from CH or CD models in the shape of the wind-sea spectrum. For the standard fetch-or duration-limited growth cases, DP models yield spectral growth curves with no overshoot effects; further, the evolution of spectrum in DP models is generally more sensitive to fetch variations with propagation direction.

3. All present second generation models suffer from limitations in the parameterization of the nonlinear energy transfer $S_{n\ell}$. Under conditions of rapidly changing wind fields, the parameterization of $S_{n\ell}$ generally contained too few degrees of freedom to cope with the wide variety of spectral distributions that may arise.

4. The variation in the basic fetch-limited growth rates of different models appear to mask the variations associated with different wind-field geometries. This may be due to uncertainties as to whether the wind speed U or the friction velocity $U*$ is the more appropriate parameter to characterize the wave growth and to the uncertainty in the precise magnitude of the drag coefficient. Another source of uncertainty is the marine boundary layer formulation for

specifying the surface winds which is the primary input for driving
the wave models; this aspect is discussed in more detail in the next
Chapter.

    5. Finally, the superposition of these differences, related
either to model class or to individual model assumption can result in
substantial net discrepancies between different model predictions for
complicated wind fields such as hurricane wind fields. Consequently,
performance of the present wave prediction models under extreme and
rapidly changing wind fields (for which direct measurements are
sparse) cannot be regarded as reliable at present.

    The SWAMP Group study has provided a closer look at some of
the recent spectral wave models and has focused on the problem of wind
specification and representation of nonlinear energy transfer in the
second generation models. However, the SWAMP study fails to provide
any insight into the performance of various wave models under operational environment.

7.3.2   Intercomparison with Real Wind Fields

    The evaluation of a wave model driven by real winds can help
bring out some of the model characteristics which may not become evident when the same model is driven under simulated wind conditions.
The increasing use of wave models in operational mode has provided an
opportunity to test and evaluate two or more wave models using real
wind fields. In an earlier section of this Chapter evaluation of a
pair of operational models belonging to CD catagory and of another
pair belonging to the CH catagory has been presented and discussed.
Another study in which three operational spectral wave models (BMO,
GONO and HYPAS) were evaluated using idealized as well as real winds
has been reported by Bouws et al. (1985). The main objective of this
study was to evaluate the parameterization of shallow-water processes
employed in the three models. Initially, the three models were numerically integrated using a constant offshore wind of 20 m s^{-1} and four
selected values of bottom depth. It was found that the models BMO and
GONO exhibited similar behaviour in the evolution of energy and peak
frequency, whereas the HYPAS model displayed less depth attenuation
and little variation in peak frequency. These three models were
further evaluated in a hindcast mode using accurately reconstructed
wind fields for the North Sea storms of 18 to 26 November 1981. The
model results were evaluated against measured data at two waverider
locations in the North Sea. The time series of wave height and period
agreed quite well with measurements, BMO and HYPAS predicting correct
energy levels except at storm peaks and GONO generally overpredicting
the energy levels during the storm period. The hindcast evaluation

results also revealed some of the shortcomings of the models namely, excessive growth rates in fetch- and duration-limited situations for GONO and an overall negative bias of energy levels for BMO and HYPAS. The authors (Bouws et al.) however conclude that the hindcast period of 18 to 26 November 1981, although carefully chosen based on storm patterns and data availability, was not adequate enough to isolate the effects of various modelling processes.

7.4 Wave Model Intercomparison During CASP

A comprehensive wave model intercomparison study was initiated by the AES (Atmospheric Environment Service) during the field project of the Canadian Atlantic Storms Program (CASP). The CASP was an intensive observational program aimed at studying the evolution of the winter storms that affect the Canadian Atlantic provinces. A large amount of meteorological and oceanographical data was collected during the CASP field project, 15 January to 15 March 1986 over the CASP area covering the Canadian Atlantic provinces and the adjoining continental shelf region. The CASP provided a unique opportunity for AES to assess the performance of its present wave analysis and forecasting procedure and also evaluate two spectral wave models which are being considered for operational implementation. As mentioned earlier (in Ch. 4), a computerized procedure based on Wilson's moving fetch technique has been developed and put into operation to prepare significant wave height charts for northwest Atlantic and northeast Pacific; this procedure is identified as the Parametric Wave Model (PWM) of the AES. Besides this, the Meteorology and Oceanography Centre (METOC) in Halifax (Nova Scotia) and Victoria (British Columbia) routinely prepare significant wave height charts based on a combination of ship data, Bretschneider nomogram and continuity of wavefield patterns. During the CASP field project, wind and wave products from the AES operational model PWM as well as wave heights from the METOC charts were evaluated against the measured wind and wave data collected over the Canadian Atlantic during the period 15 January - 15 March, 1986. The two spectral wave models that were tested during the CASP field project were the ODGP and the WAVAD. (these models are described in earlier Chapters). The three models namely PWM, ODGP and WAVAD were driven by winds provided by the weather prediction model of the Canadian Meteorological Centre (CMC) in Montreal. In addition, the ODGP model was also run in parallel with operational (OPR) winds which uses a 'man-machine mix' procedure. The two sets of wind and wave products from the ODGP model were designated as ODGP-CMC and ODGP-OPR respectively. The model products were evaluated over three deep-water regions as shown in Figure 7.8. This Figure shows the wave measuring sites namely the NOAA buoys and the offshore drilling rigs and also the grid points of ODGP, WAVAD and PWM/METOC where the model products

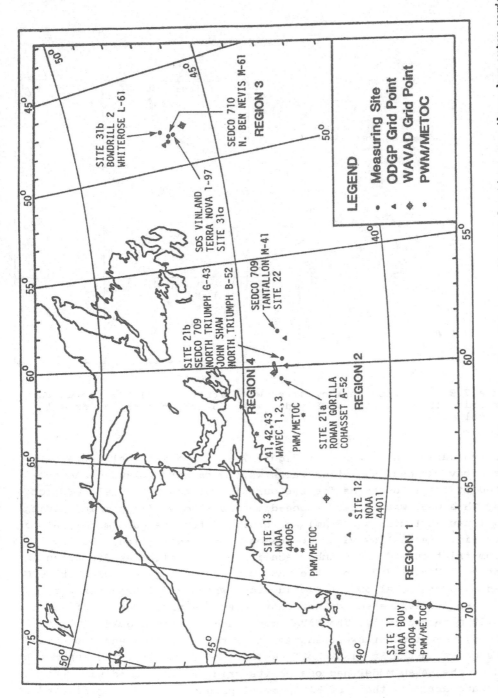

Figure 7.8: The CASP study area showing the various measuring sites and grid point locations over three deep-water regions and one shallow-water region.

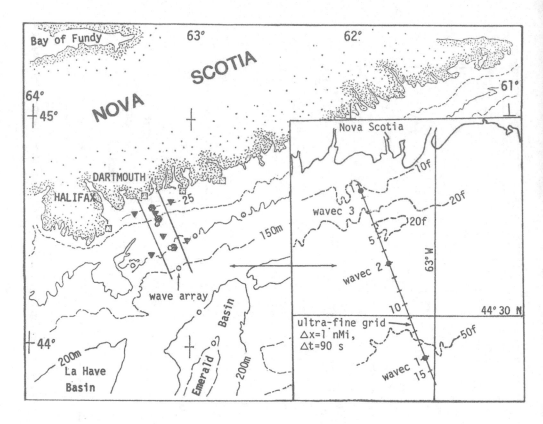

Figure 7.9: The CASP shallow-water wave array (left) and the ultra-fine grid (inset) of the ODGP model with 15 grid points and Δx = 1 nautical mile.

were evaluated. In region 4 of Figure 7.8, a special shallow-water wave array was designed along which wave data were measured and collected at three locations for the duration of the CASP field project. Along this CASP wave array, a one-dimensional grid (with grid spacing Δx = 1 nautical mile ≈ 1.8 km) was designed for the ODGP model; along this grid, the shallow-water processes namely refraction, shoaling, bottom-friction and wave number scaling were modelled as discussed in Chapter 6. Figure 7.9 shows the CASP wave array together with the ODGP one-dimensional shallow-water grid along which model products were generated and evaluated against buoy measurements at depths of 25, 50 and 100 m respectively. The WAVAD model used a coarse grid (2° lat. x 2° long) covering northwest and north central Atlantic and two nested grids covering the Scotian shelf region (see Fig. 6.8). The innermost grid of the WAVAD model had one of its grid array aligned with the CASP wave array so that the WAVAD model products could be generated at the three shallow-water buoy locations. The evaluation of various

models was done against all available data during the two month period
of the CASP field project. A large number of wind and wave plots and
scatter diagrams of observed vs. model values were prepared; also,
various error parameters like RMSE, ME, SI etc. were calculated.
Selected results of this evaluation are presented in the following
subsection.

7.4.1 Evaluation Over Deep-water Sites

For each of the three deep-water regions (Fig. 7.8) two sites
were chosen at which model products were generated and evaluated
against measured products available at the nearest location. For each
day, model products were generated at four standard meteorological
times (00, 06, 12, 18 GMT) and evaluated against observed products.
Thus, over the 60-day period (15 January - 15 March, 1986) a total of
1440 data points would be generated if there were no missing values.
There were however several missing observations and further two of the
models namely PWM and METOC generated products only twice a day; consequently the number of data points varied from a low of about 325 to
a high of over 1300 for different models. In Figure 7.10 are shown
scatter diagrams of observed vs. model wind speed and significant wave
height for the three spectral wave models namely ODGP-CMC, ODGP-OPR
and WAVAD respectively; these scatter diagrams relate to model products generated at analysis time (zero-hour forecast). In Figure 7.11
are shown similar scatter diagrams for the three spectral models at
24-hour forecast time, while Figure 7.12 shows scatter diagrams at
36-hour forecast time. The results of the two nonspectral wave models
namely PWM and WAVAD are shown in Figure 7.13 which shows the scatter
diagrams for 00-hour, 24-hour and 36-hour forecast time respectively.
For these two models (PWM and METOC) scatter diagrams are presented in
respect of significant wave height only. Besides wind speed and significant wave height, the peak period (T_p) is another important
product of a wave model which can be evaluated against measured peak
period as determined from the peak frequency of the wave spectrum. In
Figures 7.14 and 7.15 are shown scatter diagrams between observed and
model peak period at 00-hour and 24-hour forecast time in respect of
four models namely ODGP-CMC, ODGP-OPR, PWM and WAVAD respectively. The
METOC charts provided only significant wave height, hence no peak
period evaluation is available for the METOC model. All the scatter
diagrams show the number of data points (N) used, the linear correlation coefficient (r) between observed and model value and the line of
regression between observed and model values. The scatter diagrams
provide a qualitative evaluation between model and observed values.
For a quantitative evaluation, various error statistics are worked out
in respect of all the five models and are shown in Table 7.V; the
error statistics are worked out upto 36-hour forecast projection time
in respect of wind speed and significant wave height.

00-HOUR FORECAST

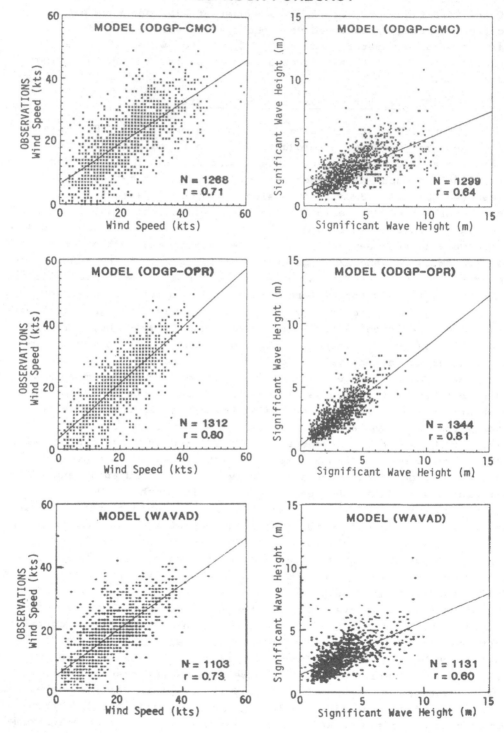

TABLE 7.V: Evaluation of wave models during CASP field project
(15 Jan.-15 Mar. 1986; all deep-water sites)

SUMMARY OF ERROR STATISTICS								
		WIND SPEED (Knots) Forecast Time				WAVE HEIGHT (Metres) Forecast Time			
Model	Parameter	00 hr	12	24	36	00 hr	12	24	36
ODGP CMC	RMSE	7.6	8.2	8.5	9.3	1.88	1.91	1.96	1.97
	MAE	5.7	6.1	6.4	7.0	1.33	1.37	1.43	1.46
	ME	0.9	2.1	2.5	2.5	1.02	1.10	1.19	1.21
	SI	36	39	41	45	65	66	69	70
	r	0.71	0.70	0.67	0.60	0.64	0.65	0.66	0.66
	N	1268	1270	1268	1260	1299	1299	1295	1285
ODGP OPR	RMSE	5.9	7.2	7.9	8.4	0.85	0.92	0.98	1.06
	MAE	4.5	5.6	6.2	6.6	0.65	0.70	0.74	0.80
	ME	-1.3	-0.5	-0.5	-0.9	0.28	0.26	0.24	0.19
	SI	28	35	38	40	30	32	34	37
	r	0.81	0.69	0.62	0.58	0.81	0.79	0.76	0.72
	N	1312	1312	1312	1312	1344	1344	1344	1344
WAVAD	RMSE	6.2	6.5	7.0	7.5	1.54	1.55	1.66	1.72
	MAE	4.6	5.0	5.5	5.9	1.10	1.12	1.20	1.26
	ME	-0.1	0.8	1.5	1.6	0.52	0.64	0.83	0.91
	SI	33	35	38	41	53	54	58	60
	r	0.73	0.71	0.69	0.64	0.60	0.63	0.65	0.64
	N	1103	1105	1104	1097	1131	1131	1128	1119
PWM	RMSE	7.0	7.8	8.3	9.2	1.26	1.37	1.54	1.58
	MAE	5.1	5.7	6.2	6.6	0.82	0.84	1.00	1.05
	ME	-2.2	-0.3	0.6	0.9	-0.20	0.01	0.21	-0.01
	SI	35	37	40	46	47	50	56	58
	r	0.81	0.76	0.72	0.65	0.69	0.68	0.64	0.57
	N	321	318	316	313	334	331	329	326
METOC	RMSE	For the METOC model, only wave height charts were evaluated, hence no error statistics for wind speed were generated.				0.83	1.20	1.15	1.23
	MAE					0.62	0.88	0.89	0.95
	ME					0.28	0.34	0.25	0.21
	SI					28	41	39	42
	r					0.83	0.61	0.62	0.56
	N					537	543	553	553

RMSE: Root Mean Square Error; MAE: Mean Absolute Error
ME: Mean Error; SI = Scatter Index
r: Linear correlation coefficient between model and observed
value; N: Number of data points

Figure 7.10: Scatter diagram between observed versus model wind speeds
(left) and between observed versus model significant wave heights
(right) for all deep-water sites (see Fig. 7.8) covering the two-month
CASP period. The scatter diagrams are shown for three spectral wave
models used in the CASP study. Also shown in each scatter diagram are
the number of data points (N), the linear correlation coefficient (r)
and the line of regression between observed and model values.

24–HOUR FORCAST

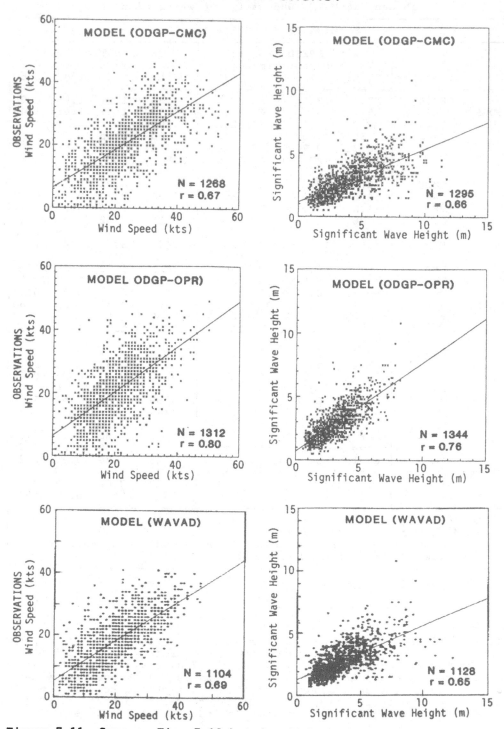

Figure 7.11: Same as Fig. 7.10 but for 24-hour forecast time.

36-HOUR FORECAST

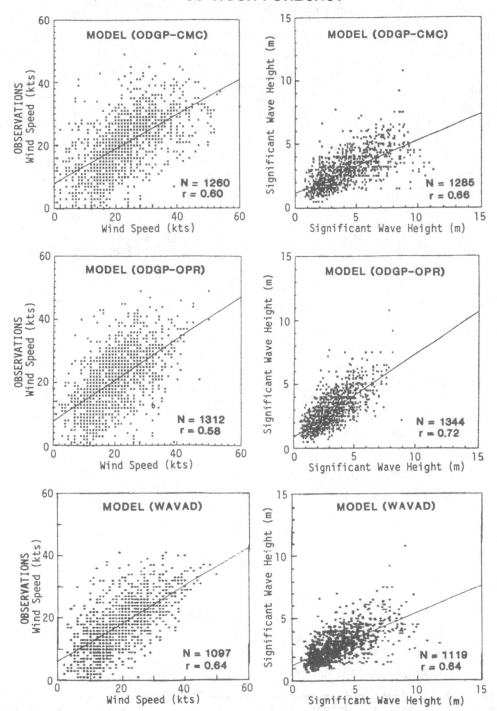

Figure 7.12: Same as Fig. 7.10 but for 36-hour forecast time.

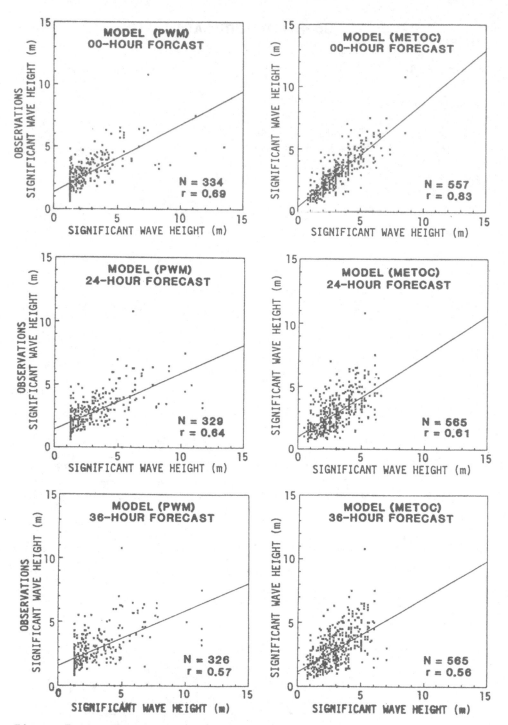

Figure 7.13: Scatter diagrams between observed versus model significant wave height at 00-hour, 24-hour and 36-hour forecast time for two
nonspectral models, namely PWM (left) and METOC (right), covering the
same two-month CASP period. Also shown are the number of data points
(N), the linear correlation coefficient (r) and the line of regression
between observed and model values.

00-HOUR FORECAST

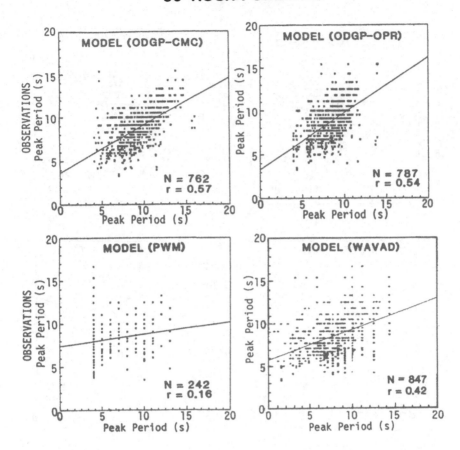

Figure 7.14: Scatter diagram between observed and model peak period (T_p) at analysis time (00-hour forecast) for all deep-water sites. The scatter diagrams are shown for four models, namely ODGP-CMC, ODGP-OPR, PWM and WAVAD respectively and cover the same two-month CASP period. Also shown are the number of data points (N), the linear correlation coefficient (r) and the line of regression between observed and model values.

24-HOUR FORCAST

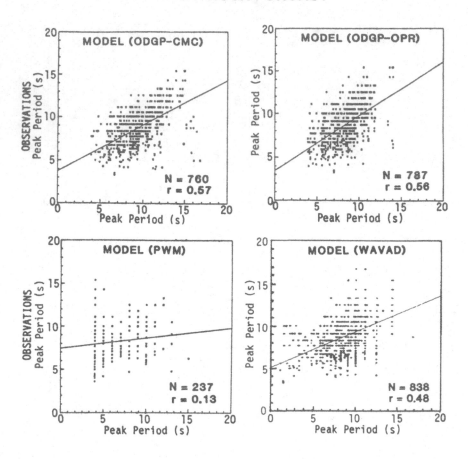

Figure 7.15: Same as Fig. 7.14 but for 24-hour forecast time.

The various scatter diagrams of Figs. 7.10 to 7.15 and the error statistics of Table 7.V reveal several interesting features of this intercomparison study. In general, it is found that model winds with better error statistics (i.e. low values of RMS error and scatter index and high values of correlation coefficient) produce better wave products which are in closer agreement with observed values. The ODGP-OPR winds, which are generated by a man-machine mix procedure, yield the best error statistics at analysis time and consequently provide the best wave products in analysis as well as in forecast mode. At analysis time, the METOC charts appear to provide the closest agreement with the observed wave heights with a RMS error of 0.83 and a correlation coefficient between observed and model values of 0.83. This closest agreement may be due to the fact the METOC charts incorporate observed wave data in their analysis. In a forecast mode, however, the ODGP-OPR model provides better results than METOC when measured in terms of RMS error, scatter index and the correlation coefficient. The METOC forecast charts are based on an empirical procedure, hence their forecast wave heights show a smaller skill than that attainable by a spectral model like ODGP which is well tuned and which is driven by winds which are better specified at analysis time than those extracted directly from a weather prediction model. This aspect of improved wind specification and its impact on model wave products will be discussed in detail in Chapter 8. The wave products for the two models ODGP-CMC and WAVAD were obtained using the same CMC winds although different interpolation techniques were used to generate winds at respective model grids. Further, the WAVAD and the ODGP grid points have different position coordinates and the sample sizes (N) for ODGP-CMC and WAVAD are significantly different; consequently, the wind error statistics for the two models (ODGP-CMC and WAVAD) show some significant differences even though both the models were driven by the same CMC wind fields. Nevertheless, the differences in the wave height error statistics can be attributed primarily to the differences in the model physics. For the 2-month CASP period, the WAVAD error statistics (RMSE, MAE and SI) show an improvement of about 15 percent over the corresponding ODGP-CMC statistics; this improvement may be attributed to the WAVAD model physics which includes nonlinear interaction terms in a parameterized form. Finally, the scatter diagrams for the peak period (Figs. 7.14 and 7.15) suggest that in general, the peak period values generated by all the four models are poorly correlated with the corresponding observed values in analysis as well as in forecast mode.

7.4.2. Evaluation over Shallow-water Sites

Along the shallow-water array (Figure 7.9) wind and wave products were generated by the two models, namely ODGP-CMC and WAVAD. As

TABLE 7.VI: Evaluation of wave models during the CASP field project (15-Jan. - 15 Mar. 1986; all shallow-water sites)

Summary of error statistics									
		Wave height (m) Forecast time				Peak Period (s) Forecast time			
Model	Parameter	00 hr	12	24	36	00 hr	12	24	36
	RMSE	0.90	0.93	1.02	1.13	2.33	2.39	2.41	2.56
	MAE	0.57	0.61	0.68	0.74	1.63	1.68	1.74	1.86
ODGP	ME	0.30	0.37	0.41	0.41	0.62	0.62	0.79	0.78
CMC	SI	55	57	63	69	26	26	27	28
	r	0.85	0.86	0.84	0.77	0.63	0.62	0.61	0.59
	N	457	460	463	463	457	460	463	463
	RMSE	0.87	0.92	0.81	0.86	4.0	3.8	3.6	3.5
	MAE	0.63	0.67	0.60	0.63	2.8	2.7	2.6	2.5
WAVAD	ME	-0.73	0.49	0.11	0.94	-1.6	-1.4	-1.2	-0.8
	SI	54	55	48	51	44	41	39	38
	r	0.70	0.68	0.72	0.69	0.37	0.43	0.42	0.46
	N	463	466	469	469	463	466	469	469

Mean observed values of parameters at individual shallow and deep water sites (00 hr) for the duration of the CASP period.

	Shallow-water sites		
	Site41	Site42	Site43
Mean wave height	1.71	1.49	1.15
Mean peak period	8.88	9.00	9.15

	Deep-water regions		
	Region 1	Region 2	Region 3
Mean wave height	2.45	2.51	3.66
Mean peak period	8.10	8.88	10.22

before, these products were generated over the two-month period of the CASP field project, at three shallow-water sites (41, 42 and 43) corresponding to the three WAVEC buoys where the water depths were 100,50 and 25 m respectively. These model products were evaluated against measured data and the various error parameters are presented in Table 7.VI. The upper half of Table 7.VI shows the various error parameters calculated using all shallow-water points while the lower half of the Table shows the mean value of the wave height and the peak period at individual shallow as well as deep-water sites. For both the models (ODGP-CMC and WAVAD), the wind specification over the shallow-water area was based on a suitable interpolation of the wind at the nearest coarse grid; accordingly, error statistics for wind speed is not presented for the shallow-water region since it reflects essentially the same skill as that over the deep-water region.

Table 7.VI reveals several interesting aspects of shallow-water wave parameters. First, the average observed values of wave

heights are smaller over the shallow-water sites by 30 to 50 percent
when compared against average wave height values over the deep-water
sites. Further, the observed mean value of the wave height steadily
decreases with decreasing depth of the observing site. As the deep-
water waves, generated during several of the CASP storms, moved toward
the CASP wave array (regions 4 of Fig. 7.8), they were generally
attenuated by shallow-water processes and this has led to smaller
values of wave heights at all shallow-water sites. These smaller
values of wave heights have further helped produce smaller values of
the error parameters (RMSE, MAE and SI) for both the models (ODGP-CMC
and WAVAD) when compared against the corresponding values at deep-
water sites (Table 7.V). In case of the peak period however, the mean
observed value has increased in general over the shallow-water sites
and consequently most of the error parameters have shown a slight
increase in magnitude when compared against the corresponding deep-
water values. Along the shallow-water array, the mean observed value
of the peak period increased steadily as the water depth decreased
indicating that longer-period swell waves generated during the CASP
storms have penetrated into shallow-waters producing higher values for
the mean peak period.

An examination of the various error parameters in Table 7.VI
suggests that the WAVAD model shows a slight improvement over the
ODGP-CMC in generating wave heights over the three shallow-water sites
taken together. However, the ODGP model appears to provide a much
better simulation of the peak period when compared against the WAVAD.
Two factors can be attributed in favour of the ODGP: 1. The ODGP
shallow-water algorithm (as discussed in Chapter 6) may have provided
a better simulation of the swell period than that of the WAVAD, which
uses wave breaking as the primary mechanism for wave height decay,
2. The WAVAD model selects the peak period based on the relative magnitude of its sea or swell energy. The WAVAD model does not generate a
single 'dominant' period based on the maximum energy density as the
ODGP model does.

The above analysis and the peak period scatter diagrams displayed in Figs. 7.14 and 7.15 suggest a need for defining a more
appropriate parameter relating wave period. A parameter 'mean wave
period' has been suggested by some modelers as a more stable measure
of the frequency distribution of the spectrum.

7.4.3 Evaluation over Selected Storm Periods

During the two months of the CASP field project, as many as 16
storm cases were identified as Intensive Observing Period (IOP). We
have chosen here two such IOP's for which the storm tracks are shown

in Figure 7.16. For these two storm periods, wind speed and wave height values were evaluated against measured data by plotting them as shown in Figures 7.17 to 7.19. In Fig. 7.17 are shown variations of wind speed and wave heights for the three spectral models (ODGP-CMC, ODGP-OPR and WAVAD) covering the storm period 15-18 February 1986; similar wind speed and wave height variation for the storm period 10-13 March 1986 is shown in Fig. 7.18. For the two nonspectral models (PWM and METOC), the wave height variation for both the storm periods is shown in Fig. 7.19. The METOC model generates only wave heights, hence wind speed variation is not presented for the two nonspectral models.

The wind speed and wave height plots of Fig. 7.17 to 7.19 reveal several interesting features associated with the passage of storm centres at different observing sites. The February storm (iden-tified as IOP 8) was developed over the Cape Hatteras region (U.S.A.), and rapidly moved into the CASP area along a typical southwest-north-east oceanic track. Associated with this storm movement is the observ-ed and the model wind variation at different sites as shown by various curves on the left side of Fig. 7.17. In association with this wind variation, wave heights of up to 5 m were first generated at site 11 (southwest of Georges Bank) and then at site 21a(over the Scotian Shelf). As the storm moved out of the CASP area, it generated high waves in the Grand Banks area (site 31b) in response to strong north-westerly winds off the Newfoundland coast. The CMC winds appear to have been overpredicted between 16th and 17 February at site 31b and this results in an overprediction of wave heights by the ODGP-CMC as well as by the WAVAD model, both of which use the CMC winds. The wave heights produced by the ODGP-OPR appear to be in better agreement with measured heights at all the three sites. For the March 10-13 storm (IOP 15), the Grand Banks area once again shows dramatic variation in wind speed as the storm centre moved from the Great Lakes area to the northeast of Newfoundland in less than two days. During the movement of the storm through the Great Lakes region on 10 March 1986, most of the CASP area was under the influence of a high pressure, providing moderate onshore (southwesterly) flow at sites 11 (Georges Bank) and 21a (Scotian Shelf), while a strong offshore (northwesterly) flow prevailed in the Grand Banks area(site 31b). In response to this flow pattern, sites 11 and 21a reported only moderate wave height values while site 31b reported wave height values of more than 5 m on 10 March 1986. As the storm centre rapidly moved through the CASP area, the Grand Banks area experienced a weak onshore flow on 11 March 1986 followed by strong offshore (northwesterly) flow when the storm centre moved south of Iceland on 12 March 1986, 1800 GMT. The significant wave height (at site 31b) increased to 7 m on 11 March, 00 GMT, then decreased to just over 1 m on 11 March, 1800 GMT and increased again

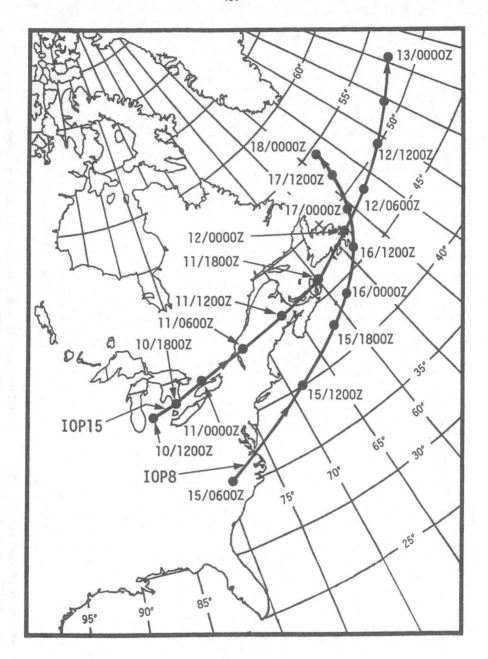

Figure 7.16: The storm tracks for two selected IOP (Intensive Observing Periods) during the CASP field project. IOP 8: 15-18 February, 1986; IOP 15: 10-13 March 1986.

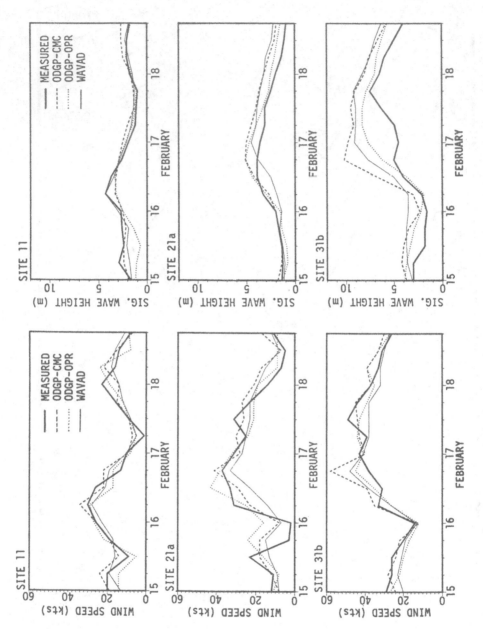

Figure 7.17: Variation of wind speed (left) and significant wave height (right) at three deep-water sites during IOP 8 for three spectral wave models used in the CASP study. For comparison the measured variation in wind speed and significant wave height is also shown

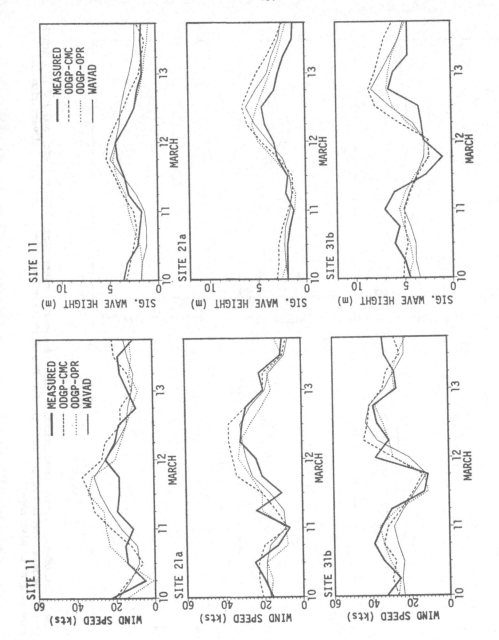

Figure 7.18: Same as Fig. 7.17 but for IOP 15 covering the period 10-13 March 1986

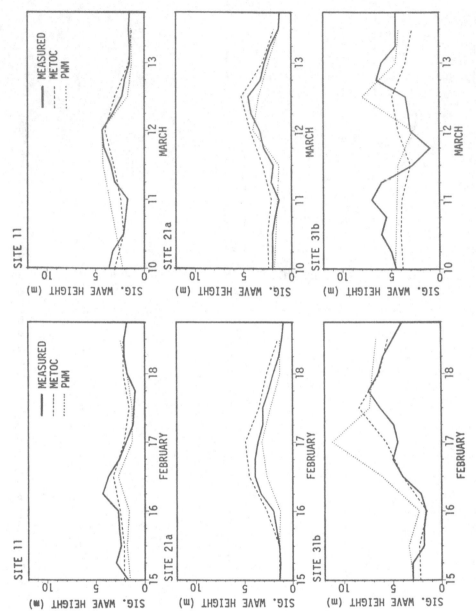

Figure 7.19: Variation of significant wave height at three deep-water sites during IOP 8 (left) and IOP 15 (right) for two nonspectral models, PWM and METOC, used in the CASP study. For comparison, the measured variation in significant wave height is also shown

to 5 m as the offshore flow was reestablished. The CMC winds were
generally overpredicted everywhere excepting at site 31b during the
first 36-hours; consequently the wave heights were slightly underpredicted at site 31b during the first 36 hours after which they were
overpredicted by both, the ODGP-CMC and the WAVAD models. The ODGP-OPR
once again appears to provide the closest agreement with measured wave
heights at all sites. Of the two nonspectral models, the METOC appears
to provide the closest agreement with measured wave height at all
sites. The PWM model, which was driven by CMC winds, overpredicted
wave heights in the Grand Banks area, most certainly due to a positive
bias in the CMC wind specification.

7.5 Summary

The various wave model validation and intercomparison studies
summarized above have provided a degree of confidence in the use of
wave prediction models and their products. Based on the various
evaluation studies reported so far, it can be concluded that a first
generation spectral wave model can produce wave height values with a
RMS error of between 1 to 1.5 m and a scatter index between 35 and
40; the same model, if properly tuned and driven with improved wind
specification can provide wave height values with a RMS error of less
than 1 m and a scatter index of 30 or less. The second and third
generation models, if driven with improved wind specification may
provide further improvements of up to 50 percent. Most of the wave
model evaluation studies have relied on limited wave measurements
available mostly along the coastal and continental shelf regions of
the world oceans. With the advent of remote-sense technology, wind and
wave measurements are becoming available over deep-water areas (along
sub-satellite tracks, for example) of world oceans. This has prompted
a few investigators (ex. Pierson, 1982; Pickett et al., 1986) to
evaluate some of the operational wave models against remote-sensed
wind and wave data derived from the satellite tracks of SEASAT and
GEOSAT. These studies have provided an increased level of confidence
in the utility of operational wave models. Hopefully, in the next
few years, satellites may provide better wind and wave database for
testing many of the wave models which have not been extensively tested
so far.

REFERENCES

Bales, S.L., W.E. Cummins and E.N. Comstock, 1982: Potential impact of
twenty year hindcast wind and wave climatology on ship design. Marine
Technology, 19, 111-139.

Barnett, T.P., 1968: Loc. cit. (Ch. 5)

Bouws, E. et al (the SWIM Group), 1985: A shallow-water intercomparison of three numerical wave prediction model (SWIM). Q. J. Royal Met. Society, 111, 1087-1112

Bretschneider, C.L., 1970: Loc. cit. (Ch. 4)

Cardone, V.J. and W.J. Pierson and E.G. Ward, 1976: Loc. cit. (Ch. 5)

Clancy, R.M., J.E. Kaitala and L.F. Zambresky, 1986: Loc. cit. (Ch. 5)

Dexter, P.E., 1974: Tests on some programmed numerical wave forecast models. J. Physical Oceanography, 4, 635-644.

Feldhausen, P.H., S.K. Chakrabarti and B.W. Wilson, 1973: Comparison of wave hindcasts at weather station J for North Atlantic storm of December 1959. Deut. Hydrogr. Zeit., 26(1), 10-16.

Günther H., G.J. Komen and W. Rosenthal, A semi-operational comparison of two parametrical wave prediction models. Deut. Hydrogr. Zeit., 37, 89-106.

Inoue, T., 1967: On the growth of the spectrum of a wind generated sea according to a modified Miles-Phillips mechanism and its application to wave forecasting. 1967: TR-67-5, Geophysical Sciences Laboratory Report, New York Univ., School of Engineering Science.

Janssen, P.A.E.M., G.J. Komen and W.J.P. De Voogt, 1984: Loc. cit. (Ch. 5).

MacLaren Plansearch Ltd. 1985: Evaluation of the spectral Ocean Wave Model (SOWM) for supporting real-time wave forecasting in the Canadian east coast offshore. Report prepared for Atmospheric Environment Service, Downsview, Ontario, MacLaren Plansearch Ltd., Halifax, Nova Scotia, 1985, 270 pp.

Pickett, R.L., D.L. Burns and R.D. Broome, 1986: Comparison of wind and wave model with GEOSAT: Final Report. Report 201, Naval Ocean Research and Development Activity, NSTL, Mississippi, U.S.A., October 1986, 8 pp.

Pierson, W.J., 1982: Loc. cit. (Ch. 5)

Pierson, W.J. and R.E. Salfi, 1979: A brief summary of verification results for the spectral Ocean Wave Model (SOWM) by means of wave height measurements obtained by GEOS-3 J. Geophysical Research, 84, B8, 4029-4040.

The SWAMP Group, 1985: Loc. cit. (Ch. 5)

Wilson, B.W., 1955: Loc. cit. (Ch. 4)

Zambresky, L.F., 1989: A verification study of the global WAM model, December 1987-November 1988. European Centre for Medium Range Weather Forecasts (ECMWF), Technical Report No. 63, Shinfield Park, Reading, U.K. (in print).

CHAPTER 8
WIND SPECIFICATION FOR WAVE ANALYSIS AND PREDICTION

Wind is the only driving force for all wave models and hence
a wave model can only be as good as the wind field that drives the
model. Ideally, wind specification must be such as to allow the important physical processes of wave generation, growth and dissipation to
be appropriately represented in a wave model. In earlier discussions,
three processes of the wave generation and growth were identified.
These are: Phillips' resonance mechanism, Miles' shear flow mechanism
and Hasselmann's nonlinear wave-wave interaction mechanism. Further,
according to recent studies, wave breaking is considered to be the
most important dissipation mechanism over open ocean. In order for
these processes to be suitably represented, wind specification on a
meso- to micro-scale would be required over an ocean basin. Such a
wind specification is rather ideal and cannot be attained on a routine
basis at present.

In early days of wave analysis and prediction, appropriate
wind information was extracted from synoptic weather maps. For example, the pioneering study of Sverdrup and Munk and several others that
followed made use of surface winds extracted from sea-level weather
charts. These sea-level weather charts in the late 1940's and early
1950's were constructed using available ship data and assuming continuity of weather patterns over oceanic regions. With the advent of
numerical weather prediction and computerized analysis procedure,
regional and hemispheric weather charts have been prepared using an
objective analysis scheme (ex., Cressman, 1959) for the last twenty
years or more. Such an objective analysis scheme provides values of
meteorological variables like wind or pressure on a grid mesh by
analyzing available observations in a given area, objectively; the
effect of atmospheric stability is, in general, not included in such
an objective analysis scheme. The landmark study of Cardone(1969)
considered atmospheric stability in terms of boundary layer processes
and developed an appropriate wind specification procedure for ocean
wave models; this study is briefly described in the following section.

8.1 Cardone's Procedure

Cardone(1969) developed a two-layer model of the Marine Boundary Layer (MBL) that includes the effects of atmospheric stability,

baroclinity and a realistic description of the lower boundary. The
atmospheric stability as measured by the temperature difference
between the sea surface and the overlying air has been identified as
an important factor influencing wave growth. Studies by Roll(1952) and
Fleagle(1956) have concluded that significantly higher waves are generated in unstable conditions (water temperature higher than the overlying air temperature by 6°C or more) than in stable conditions. These
studies, together with reliable observations of experienced mariners
have clearly demonstrated the effect of atmospheric stability on wave
growth. In order to consider the effect of atmospheric stability on
boundary layer wind flow, Cardone divided the MBL into two layers, a
surface layer and an Ekman layer overlying the surface layer. Earlier
studies, notably by Blackadar(1965) considered a two-layer representation of the atmospheric boundary layer for neutral stratification.
Cardone extended Blackadar's two-layer model to consider non-neutral
stratification by applying the Monin-Obukhov similarity theory to the
surface layer where the existence of universal relationships between
non-dimensional wind shear and temperature gradients are postulated;
these relationships can be expressed as;

$$\phi_u = \frac{kZ}{U_*} \frac{\partial U}{\partial Z} \; ; \quad \phi_t = \frac{Z}{T*} \frac{\partial \theta}{\partial Z}$$

$$(8.1)$$

$$T* = -\frac{1}{kU_*} \frac{H}{c_p \rho_a}$$

The non-dimensional gradients ϕ_u and ϕ_t are related by

$$\phi_u = \alpha_h \phi_t \quad \text{where} \quad \alpha_h = \frac{K_h}{K_m} \frac{\text{(eddy diffusivity)}}{\text{(eddy viscosity)}}$$

In (8.1), k is the von Karman constant, θ is the potential temperature, H is the sensible heat flux from sea to air, c_p is the specific
heat of air at constant pressure and ρ_a is the air density; the
symbols U and U_* are already defined earlier.

   The assumption of similarity of wind and temperature profiles
in the surface layer implies that α_h is constant and hence a modified
form of the stability length L' (also called Monin-Obukhov length;
Monin and Obukhov, 1954; Monin, 1970) can be written as,

$$L' = \frac{U_*(\partial U/\partial Z)T}{kg(\partial \theta/\partial Z)}$$

$$(8.3)$$

In (8.3), T is the air temperature (°K) and other symbols are already
defined.

The set of equations (8.1) can be integrated to obtain following equations:

$$U_Z = \frac{U_*}{k}[\ell n(Z/Z_o) - \psi(Z/L')]$$

(8.4)

$$\theta_Z - \theta_o = \frac{T^*}{\alpha_h}[\ell n(Z/Z_o) - \psi(Z/L')]$$

In the above equations, $\psi(Z/L') = \int_o^\xi \frac{1 - \phi_u(\xi)}{\xi} d\xi$ and Z_o is the roughness parameter. The expressions chosen for the stability functions are;

$$\phi_u(o) = 1 \qquad\qquad\qquad\qquad\qquad \text{Neutral}$$

$$\phi_u = 1 + \beta(Z/L'), \quad \beta = 7 \qquad\qquad \text{Stable}$$

$$\phi_u^4 - \Gamma(Z/L')\phi_u^3 - 1 = 0, \qquad \Gamma = 18 \qquad \text{Unstable}$$

Given a measurement of wind at a specific height and a measurement of temperature difference between air and the sea surface (strictly, the virtual potential temperature difference between air and the sea surface), the wind profile parameters U_* and L' can be calculated from the following equations:

$$L' = \frac{U_*^2 \bar{\theta} [\ell n(Z_a/Z_o) - \Psi(Z_a/L')]}{k^2 g(\theta_a - \theta_s)}$$

(8.5)

$$U_* = kU_m [\ell n(Z_m/Z_o) - \Psi(Z_m/L')]$$

In (8.5), $\bar{\theta}$ is the mean potential temperature in the marine boundary layer, θ_a and θ_s are the potential temperatures for air and sea surface respectively, Z_a is the height at which θ_a the potential temperature for air is determined and Z_m is the height at which the wind speed U_m is measured. Further, Z_o denotes the roughness parameter which, according to Charnock's (1955) empirical formula can be expressed as;

$$Z_o = \frac{b U_*^2}{g}$$

(8.6)

In (8.6), b is called the Charnock's constant for which a value of 0.0156 has been determined by Wu (1969). The roughness parameter has also been expressed by a relationship given by;

$$Z_o = \frac{A}{U_*^2} + BU_*^2 + C$$

(8.7)

Here A, B and C are empirically determined constants (see for example, Arya, 1977).

Equations (8.5) through (8.7) provide the wind profile solution in the surface layer. The height of the surface layer is assumed to be given by a formula used by Blackadar (1962) as;

$$h = U_g D/f \qquad (8.8)$$

Here U_g is the geostrophic wind, f is the Coriolis parameter ($f = 2\Omega\sin\phi$; Ω: earth's rotation rate and ϕ: latitude angle) and D is an empirical constant which is assigned a value of 2.7×10^{-4}. Above the surface layer h, the atmospheric flow is governed by the Ekman solution of the equation of motion in which K_m the eddy viscosity is assumed constant throughout the Ekman layer. This assumption leads to the well-known Ekman spiral solution (see for example, Holton, 1979) which is expressed as;

$$u = u_g(1 - e^{-aZ}\cos aZ) \; ; \qquad v = u_g(e^{-aZ}\sin aZ) \qquad (8.9)$$

Here u and v are the x and y components of the wind in the Ekman layer, u_g is the x-component of the geostropic wind vector and $a = \sqrt{f/2K_m}$. The geostrophic wind components u_g and v_g are expressed in standard notations as;

$$u_g = -\frac{1}{\rho f}\frac{\partial p}{\partial y} \quad ; \qquad v_g = \frac{1}{\rho f}\frac{\partial p}{\partial x} \qquad (8.10)$$

In (8.10), ρ is the air density, f is the Coriolis parameter and $\frac{\partial p}{\partial x}$, $\frac{\partial p}{\partial y}$ are atmospheric pressure gradients in x and y direction respectively. In deriving (8.9) it is assumed that the geostrophic wind is independent of height and that the atmospheric flow is oriented so that the geostrophic wind is entirely in the zonal direction ($v_g = 0$).

Cardone obtained the wind profile solution throughout the two layers by 'patching' the Ekman layer solution to the surface layer similarity solution by imposing the continuity of eddy viscosity, wind, wind shear and stress across z = h, the height of the surface layer. In order to obtain the complete solution, two more quantities were defined: the non-dimensional thermal wind vector and the angle between the surface wind vector and the surface geostrophic wind vector, known as the inflow angle. A system of equations is obtained which can be solved efficiently by the method of inverse interpolation as outlined by Cardone(1978). This procedure yields a solution of U_* which together with the corresponding values of Z_0 and L', can specify the entire wind profile below the level h. Further, this procedure can be applied to a geostrophic wind field extracted from a sea-level pressure chart and using the temperature fields for air and sea-surface, one can generate values of wind components at 19.5 m and friction velocity U_*; these are the input parameters required to drive a PTB type spectral wave model.

Cardone further provdies examples of reduction of measured winds to effective neutral wind speed at 20 m (65 ft) for several representative anemometer heights and (air-sea surface) temperature differences ($T_a - T_s$). An effective neutral wind at 20 m is defined as the wind speed which in a neutral atmosphere would produce the same surface stress as the actual wind at the given height. In Table 7 are given values of effective neutral wind speeds at 20 m level corresponding to measured wind speeds of 20, 40 and 60 knots for various (air-sea surface) temperature values. It can be seen from this table that for the same (air-sea surface) temperature difference, the wind adjustment is more when the difference is positive (stable atmosphere) than when the difference is negative (unstable atmosphere). However, for temperature difference of -8°C or more, wind adjustment is significant indicating a significant change in wave growth under unstable conditions.

TABLE 8.I.   Effective neutral 20 m wind speeds corresponding to measured wind speeds of 20, 40 and 60 knots for the indicated anemometer heights and (air-sea surface) temperature difference.

Anemometer Height (m)	$T_a - T_s$ °C	Measured 20	Wind 40	Speed (knots) 60
80	-8	21.1	38.9	55.8
"	-4	20.3	37.7	54.4
"	0	17.9	35.2	52.1
"	+2	10.6	30.7	49.2
60	-8	21.3	39.4	56.7
"	-4	20.5	38.3	55.4
"	0	18.3	36.1	53.5
"	+2	12.0	32.8	51.4
40	-8	21.5	40.1	58.2
"	-4	20.8	39.1	57.1
"	0	18.9	37.4	55.7
"	+2	14.3	35.2	54.3
20	-8	22.1	41.8	61.5
"	-4	21.4	41.0	60.8
"	0	20.0	40.0	60.0
"	+2	17.8	38.9	59.2

8.2   Wind Specification for Operational Wave Models

Cardone's procedure to specify the wind profile in the MBL has been an integral part of the U.S. Navy's operational running of the Spectral Ocean Wave Model (SOWM). As mentioned earlier, the wind

specification procedure of Cardone provided wind components at 19.5 m
level and the friction velocity (U_*) as input parameters to the
SOWM. Spectral wave models using the P-M spectrum as the limiting
spectrum require wind input at 19.5 m (64 ft) height above the ocean
surface. The P-M spectrum was formulated following a detailed analysis
of 420 wave records from two British Weather Ships in the north
Atlantic. These wave records were analyzed with reference to winds
measured by the Ships' anemometers located at 64 ft (19.5 m) level. A
similarity theory of Kitaigorodskii(1962) was used to prescribe the
P-M spectrum with two constants α and β (see eq. 4.19) and a wind
speed dependence in terms of an exponential expression. Thus the P-M
spectrum is intimately related to the wind speed at 19.5 m level and
consequently spectral wave models based on the P-M spectrum formulation require wind specification at 19.5 m level.

It may be noted that since Cardone's 1969 study, weather
prediction models have been developed in many parts of the world which
include many of the boundary layer processes explicity; consequently,
boundary layer products (winds, temperature etc.) obtainable from a
weather prediction model can be used, with minor modifications, as
inputs to an operational wave prediction model. For example, the U.S.
Navy's present Global spectral Ocean Wave Model (GSOWM) uses surface
winds and the friction velocity (U_*) values obtainable from the U.S.
Navy's Operational Global Atmospheric Prediction System called NOGAPS.
This system incorporates a parameterization that relates analyzed and
forecast synoptic-scale variables produced by the NOGAPS to the small-
scale variables that determine the turbulent-flow regime in the surface contact layer; this allows the mean vertical profiles of wind,
temperature and moisture to be computed as a function of the stability
length L'. The final products of the procedure are the winds at 19.5 m
level which contain an integration constant to increase or decrease
the wind speed depending upon the (air-sea surface) temperature difference. Thus the GSOWM is driven by the effective neutral winds at
19.5 m level. The operational wave prediction model of the British
Meteorological Office (at Bracknell, United Kingdom) used winds at
900 mb level to generate surface winds by the following formulae:

$$\text{wind speed} \qquad U = a\, U_{900}^2 + b\, U_{900} + c$$

$$\text{wind direction} \qquad \theta = \theta_{900} + d \tag{8.11}$$

In (8.11), a, b, c and d are constants determined empirically so as to
include the effect of atmospheric stability on wave growth. Equation
(8.11) provides wind speed and direction at 19.5 m level. A recent

modification allows winds at 10 m level to be used; these 10 m level winds are extracted from the operational weather prediction model of the British Meteorological Office. The HYPA model in West Germany and the GONO model in the Netherlands are both driven by winds applicable at 10 m level. The third generation WAM model uses winds applicable at 10 m level and transforms the wind speed to friction velocity U_* by eq.(3.8) which can be re-written as,

$$U_* = U\sqrt{C_D(U)} \qquad\qquad (8.12)$$

Here the drag coefficient C_D is expressed as a function of wind speed and varies according to the following formula first proposed by Wu(1980):

$$C_D(U) = \begin{cases} 1.2875 \times 10^{-3} & \text{for } U < 7.5 \text{m s}^{-1} \\ (0.8 + 0.065U) \times 10^{-3} & \text{for } U \geq 7.5 \text{m s}^{-1} \end{cases} \qquad (8.13)$$

With eq.(8.12) and (8.13), the wind input source function can be expressed in terms of U_* scaling which, according to recent studies, is a more appropriate approach to characterize the wave growth.

8.3 Examples of Wind Specification Differences; Results from CASP

As mentioned earlier (section 7.5), the ODGP model was driven using two different wind fields during the CASP field project. These two wind fields were designated as CMC and OPR. The CMC wind field refers to the winds extracted from the lowest active level of the weather prediction model at the Canadian Meteorological Centre, Montreal. The lowest active level is presently set at $\sigma = 0.998$ where σ is the ratio (p/p_s), p being the pressure at the level in question and p_s is the sea-level pressure. In a standard atmosphere, the level $\sigma = 0.998$ corresponds to approximately 17 m above the sea-level. The OPR winds refer to the operational winds generated by a 'man-machine mix' procedure. In this procedure, the six-hourly north Atlantic surface charts prepared and distributed by the National Meteorological Centre (NMC) in Washington, D.C. (U.S.A) are reanalyzed based on latest ship weather reports and then digitized. The boundary layer model of Cardone(1969,1978) is then applied to generate effective neutral winds at 20 m level above the ocean surface; these winds are then identified as the OPR winds for the ODGP model at analysis time. For forecast winds, the prognostic surface pressure charts generated by the Nested Grid Model at NMC are used; these forecast charts are modified through forecaster intervention in an attempt to remove systematic errors in specification of cyclone and anticyclone central pressure and corresponding pressure gradients. Errors due to

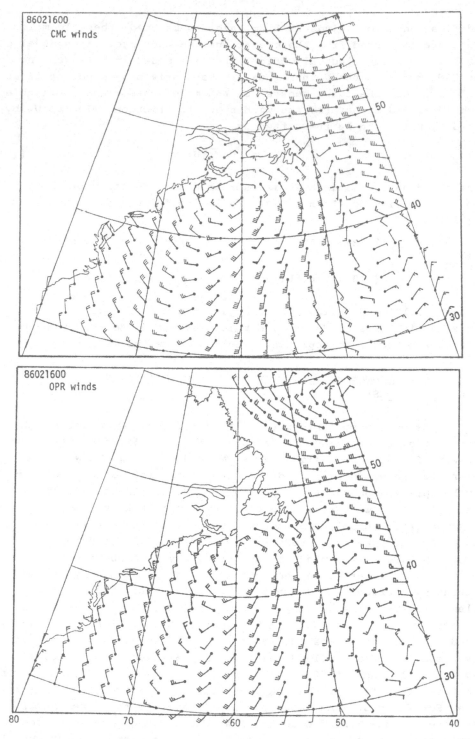

Figure 8.1: Surface wind field over the north Atlantic for 16 February 1986, OOGMT. Top: CMC model Bottom: OPR 'man- machine mix' model (wind speed in knots).

initialization and persistence are also accounted for at this step. Following this step, the boundary-layer model is applied to generate effective neutral winds at 20 m level; this provides the forecast 'man-machine mix' or OPR winds. In Figure 8.1(a,b) are shown the two wind fields namely CMC and OPR for 16 February 1986, 00Z. The upper half of Figure 8.1 is the CMC wind field over the north Atlantic while the lower half (Fig. 8.1b) shows the corresponding OPR wind field. The date (16 February 1986) corresponds to one of the 15 Intensive Observing Periods (IOP) of the CASP field project. A general inspection of Figure 8.1 shows significant differences in the two wind fields. In order to appreciate these differences quantitatively, scalar and vector differences between the two wind fields are calculated at every grid point and Figure 8.2 shows these differences.

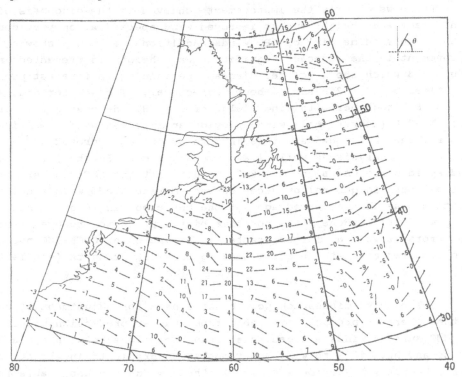

Fig. 8.2: Scaler and vector differences between CMC and OPR wind fields of Fig. 8.1. The vector represents the difference in wind direction (see inset), while the number indicates the difference in wind speed (knots) between the two wind fields.

The vectors in Fig. 8.2 represent the difference in wind direction while the number at the end of each vector represents the difference in the wind speed. A vector parallel to the latitude lines indicates no difference in wind direction between the two wind fields. A positive number at the end of the vector indicates that the CMC wind speed is greater than the corresponding OPR wind speed. Figure 8.2 clearly shows the spatial differences in the two wind fields, particularly in the vicinity of the low pressure centre southwest of Newfoundland where the differences in the wind speed and direction can be as large as 20 knots and 45 degrees respectively. Elsewhere, outside of the low pressure area, the differences in wind speed and direction are still large in some areas, more so in wind speed than in wind direction. The impact of these wind specification differences on wave model output can be clearly seen in some of the CASP results presented earlier in Chapter 7. For example, Fig. 7.10 shows scatter diagrams between observed and model products pertaining to all deep-water sites for the duration (15 Jan-15 Mar. 1986) of the CASP field project. An inspection of the scatter diagrams for ODGP-CMC vs. ODGP-OPR shows clearly the improvement achieved in the diagnosis of significant wave height due to improved wind specification provided by the 'man-machine mix' or OPR winds. Additional evidence showing improvement in the forecast significant wave height is presented in Figure 8.3 which shows scatter diagrams pertaining to forecast projection times of 12-, 24- and 36-hour respectively. The scatter diagrams in Fig. 8.3 together with wind fields of Fig. 8.1 demonstrate clearly that the OPR (operational) winds provide an improved wind specification for the ODGP model and this in turn generates improved wave products at analysis as well as at forecast times. Furthermore, the wind speed differences in Fig. 8.2 suggest that the CMC weather prediction model in general produces surface marine winds with a definite positive bias when compared against the corresponding operational (OPR) winds of Cardone; this positive bias appears to produce larger RMS errors as well as larger scatter index vaues for ODGP-CMC products when compared against the corresponding ODGP-OPR products (see Table 7.V).

The above discussion points out a need for adjustment of some of the numerical products available from weather prediction models. The CMC surface winds are generated using relations between surface stress and heat flux to wind and temperature gradients in the atmospheric boundary layer (Delage, 1985); these winds are applicable to a variety of purposes, ex. ocean wave model, storm surge model, oil spill trajectory model etc. The man-machine mix procedure of Cardone starts with surface pressure and produces OPR winds applicable specifically for driving an ocean wave model. Consequently, the CMC winds need to be 'fine-tuned' before being used for driving a spectral wave model. Similar fine tuning may be required in respect of winds extracted from other weather prediction models.

Besides the techniques discussed above, there are other procedures available which can generate marine winds for driving an ocean

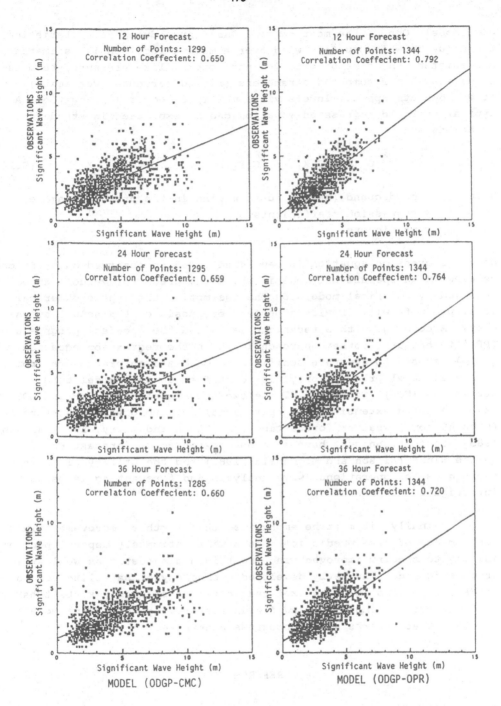

Figure 8.3: Scatter diagrams between observed versus model significant wave height, at 12-, 24- and 36-hour forecast projection times.
Left: wave heights from model ODGP-CMC; Right: wave heights from model
ODGP-OPR (from Khandekar and Eid, 1987; Copyright by American Society
of Civil Engineers)

wave model. One of the most commonly used procedures is a statistical technique in which surface wind over a marine location is estimated as a predictand in terms of observed parameters (like pressure, temperature) as well as computed parameters (like divergence, vorticity) at various atmospheric levels via a multiple regression equation. A typical multiple regression equation can be expressed in standard notations as;

$$Y = C_o + C_1X_1 + C_2X_2 \ldots \ldots \ldots \ldots + C_nX_n \qquad (8.14)$$

Here Y = predictand (marine surface wind in the present example)

C_n = regression coefficients

X_n = predictors

Once a regression equation is developed based on observed data, it can be used in an operational mode in which the predictor values can be supplied by numerical models on the assumption that these numerical model products will provide a perfect prognosis of the actual atmosphere. Accordingly, this technique is called the 'perfect prognosis (PP)' technique. Another approach in which the regression equation (8.14) is used to generate coefficients C_n based on predictors from numerical model products only, is identified as the 'model output statistics (MOS)' technique. These two techniques namely PP and MOS have been used extensively in operational meteorology for short range (0 to 48 hour) weather forecasting (ex. Glahn and Lowry, 1972; Wilson, 1985). These techniques have also been extented to forecasting of marine winds (Wilson and MacDonald, 1985) and the utility of these techniques for operational wave analysis and forecasting is being investigated at present.

Finally, it must be emphasized that further improvement in the performance of wave prediction models will intimately depend upon our ability to specify improved wind input in a diagnostic as well as in a prognostic mode. The newly developed technology of satellite-sensed winds over oceans is being explored in recent studies (ex. Lalbeharry, 1988) to provide improved wind fields for driving ocean wave models; similar other efforts are in progress elsewhere.

REFERENCES

Arya, S.P.S. 1977: Suggested revisions to certain boundary layer parameterization schemes used in atmospheric circulation models. Monthly Weather Review, 105, 215-227.

Blackadar, A.K., 1962: The vertical distribution of wind and turbulent exchange in a neutral atmosphere. J. Geophysical Research, 67, 3095-3102

Blackadar, A.K., 1965: A simplified two-layer model of the baroclinic neutral atmospheric boundary layer. Air Force Cambridge Res. Laboratory, Massachussetts, U.S.A. Rept. 65-531, 49-65

Cardone, V.J., 1969: Loc. cit. (Ch. 1)

Cardone, V.J., 1978: Specification and prediction of the vector wind on the United States continental shelf for application to an oil trajectory forecast program. Final Report, Techniques Development Laboratory, NOAA, Silver Spring, Maryland, U.S.A.

Charnock, H., 1955: Wind stress on a water surface. Q.J. Royal Meteor. Society, 81, 639-640.

Cressman, G.P., 1959: An operational objective analysis scheme. Monthly Weather Review, 87, 367-374

Delage, Y., 1985: Surface turbulent flux formulation in stable conditions for atmospheric circulation models. Monthly Weather Review, 113, 89-98.

Fleagle, R.G., 1956: Note on effect of air-sea temperature difference on wave generation. Tran. American Geophysical Union, 137, 275-277.

Glahn, H.R. and D.A. Lowry, 1972: The use of model output statistics (MOS) in objective weather forecasting. J. Applied Meteorology, 11, 1203-1211.

Holton, J.H., 1979: An introduction to dynamic meteorology, second edition. Academic Press, International Geophysical series. Vol. 23, 391 pp.

Khandekar, M.L. and B.M. Eid, 1987: Wind specification for spectral ocean wave models. Proc. 20th coastal Engineering conference, Taipei, Taiwan, 9-14 November 1986; American Society of Civil Engineers, Ch. 28, 354-365.

Kitaigorodskii, S., 1962: Loc. cit. (Ch. 5)

Lalbeharry, R., 1988: Application of SEASAT scatterometer winds for ocean wave analysis and modelling. Internal Report, MSRB-88-3, Atmospheric Environment Service, Downsview, Ontario, Canada. January 1988, 98 pp.

Monin, A.S., 1970: The atmospheric boundary layer. Annual Review of Fluid Mechanics, 2, 225-250.

"   " and A.M. Obukhov, 1954: Basic laws of turbulent mixing in the ground layer of the atmosphere. Akad. Nauk. SSSR, Geofiz. Inst. Trudy 151, 163-187.

Roll, H., 1952: Uber Grössenunterschiede der Meereswellen bei warm und kaltluft. Deut. Hydrogr. Zeit. 5, 111-114.

Wilson, L.J., 1985: Application of statistical methods to short range operational weather forecasting. Preprints, Ninth conf. on Probability and statistics in Atmospheric Sciences, Virginia Beach, American Meteor. Society, 1-10.

"   "   "   " and K. MacDonald, 1985: Assessment of perfect prog marine wind forecasts. Pro. International Workshop on offshore winds and icing, Halifax Nova Scotia, Atmospheric Environment Service, Downsview, Ontario, 352-363.

Wu, J. 1980: Wind-stress coefficient over sea surface near neutral conditions-a revisit. J. Physical Oceanography, 10, 727-740.

Wu, J. 1969: Wind stress and surface roughness at air-sea interface. J. Geophysical Research, 74, 444-455.

CHAPTER 9
WAVE ANALYSIS: OPERATIONS AND APPLICATIONS

9.1 General Comments

Increasing marine and offshore activities in recent years has
created a need for more and improved knowledge of the state of the sea
either at a given location or over a given area. The term sea-state
forecasting refers to forecasting of waves and swells over the sea.
According to Britton(1981), sea-state forecasting is an operational
problem requiring the use of practical relationships between the
atmosphere and the ocean. The Bretschneider nomogram (Fig. 4.4) may be
considered as one such practical relationship which yields the sea-
state information when the atmospheric parameters namely wind speed,
wind duration and over-water wind fetch are given. The Beaufort wind
scale (Table 4.I) with its associated wave height values and sea-state
photographs (Allen,1983) may be considered as another example of a
practical relationship and a visual guideline to determine the sea-
state information.

Before attempting to develop suitable techniques for wave
analysis, operations and applications, it is essential to develop a
firm observational basis of the sea-state at a given location as well
as over a given nearshore region. The quantity which best defines the
sea-state is E, the amount of energy required to create a deformed sea
surface. The quantity E cannot be measured directly and has to be
estimated or derived from measurements of wave height, period and
length. In the past, these wave parameters have been estimated visually by experienced observers and even to-day most of the wave and swell
observations along major shipping routes are done visually. However,
in the coastal and nearshore regions, instrument based wave observations have been made since the last twenty-five years or more and at
present a network of wave recording stations has been established
along various coastlines of north America, Europe and elsewhere. One
of the most commonly used instrument is a waverider which is an
inertial instrument as it follows the sea surface providing wave
height measurements by twice integrating the acceleration signal of a
vertically stabilized platform (buoy) in time and in some cases, also
measuring the buoy tilt in two orthogonal directions. Such a waverider
is generally mounted in moored buoys or in stable platforms such as an
oil rig. A waverider produces a wave record as shown in figure 4.1a.

Typically, a wave recorder is designed to produce a wave record of twenty minutes duration every three hours. A careful analysis and processing of such a wave record can produce a wealth of information as detailed in the following sections.

9.2 Analysis of Wave Records

A wave record may be looked upon as a time series showing oscillations of water surface about a mean level. Such a time series can be analyzed in the frequency domain by Fourier transforming the time domain signal. This results in a statistical distribution E(f) as a function of the frequency f, which is called a variance spectrum or a power spectrum. A power spectrum analysis of ocean wave records was first described by Pierson and Marks (1952). With the recent development of a Fast Fourier Transform (FFT), the power spectrum analysis of an ocean wave record can be performed with sufficient resolution in a short time. For example, the Marine Environmental Data Service (MEDS) of the Department of Fisheries and Oceans in Ottawa, Canada, uses approximately 60 discrete values of frequency between 0.05 and 0.5 Hz to analyze the wave records using the Cooley-Tukey (1965) FFT algorithm. This procedure allows the wave record of Figure 4.1a to be transformed into a spectrum plot as shown in Figure 9.1.

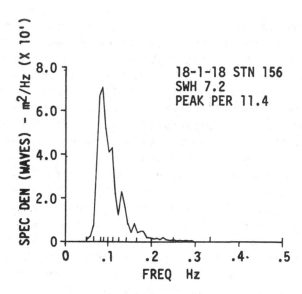

Figure 9.1: A frequency plot for the wave record of Fig. 4.1a. obtained using the Cooley-Tukey FFT algorithm. The ordinate is spectral density (m²/Hz), while the abscissa represents frequency in Hz. The significant wave height (SWH) and the peak period are also shown. The wave record was collected at Ocean ranger (station 156), east of Newfoundland in the Canadian Atlantic on 1 April 1981 (from archives of MEDS, Ottawa).

Here, the ordinate shows the spectral density in m^2/Hz, while the abscissa shows the frequency in Hz. For the spectrum plot of Fig. 9.1, the peak period is 11.4 s which corresponds to the peak frequency of about 0.09. The most important property of the variance spectrum is that the total area under the spectral curve gives m, the variance of the sea-level displacement; thus,

$$\int_o^\infty E(f)df = m \qquad (9.1)$$

It may be recalled that m is the only parameter associated with the Rayleigh distribution (see eq. 4.9) which is used to describe the distribution of wave heights H. Knowing the parameter m, the significant wave height is given by (eq. 4.11).

$$H_s = 4\sqrt{m} \qquad (9.2)$$

In the spectral plot of Fig. 9.1, the significant wave height comes out to be 7.2 m and this value is displayed as SWH in the upper right hand corner of the Figure.

The use of Rayleigh distribution allows us to compute other wave parameters that have practical applications. Consider once again the eq. (4.9) which defines the probability density function for the wave height. The expression p(H)dH is interpreted as the probability of a wave height being in the interval from H-dH/2 to H + dH/2 for all H and for dH however small. If eq. (4.9) is integrated from O to H, we have.

$$p(H) = \int_o^H p(H)dH = \int_o^H \frac{H}{4m} \exp(-H^2/4m)dH \qquad (9.3)$$

Here p(H) is the probability that a wave height is equal to or less than H. Next, if we denote $P(H_1)$ as the probability that a wave height does not exceed a value H_1, then we can write,

$$P(H_1) = 1 - \int_{H_1}^\infty \frac{H}{4m} \exp(- H^2/4m)dH$$

If we substitute $H_s = 4\sqrt{m}$ and simplify the integral, we obtain

$$P(H_1) = 1 - \exp[- 2 (H_1/H_s)^2] \qquad (9.4)$$

Equation (9.4) gives the probability that a wave height does not exceed a given value H_1; in (9.4), H_S is the significant wave height. Further, if H_S is to be computed from a wave record of finite length, the record length or the number of waves used for the computations should be taken into account. For example, on a wave record containing N waves, if n (n ≤ N) waves exceed a given height H_1, the observed probability of wave heights exceeding H_1 will be given by,

$$p(H_1) = \frac{n}{N}$$

$$\text{or} \quad P(H_1) = 1 - \frac{n}{N} \tag{9.5}$$

combine (9.5) with (9.4) to yeild

$$H_s = \frac{H_1}{\sqrt{\frac{1}{2} \ln \frac{N}{n}}} \tag{9.6}$$

Equation (9.6) provides a quick method of determining the significant wave height H_s from a given wave record. For n = 1, Equation (9.6) presents a special case; it refers to the probability of the height of the highest wave on a record containing N waves. If we denote the probable maximum height by H_{max}, we can simplify (9.6) using n = 1 and obtain

$$H_{max} = H_s \sqrt{\frac{\ln N}{2}} \tag{9.7}$$

Equation (9.7) gives the probable maximum height in a record containing N waves. A recent statistical analysis by Forristall(1978) yields an expression,

$$H_{max} = H_s \sqrt{\frac{\ln N}{2}} \left(1 + \frac{\gamma}{2 \ln N}\right) \tag{9.8}$$

In (9.8), γ is Euler's constant whose value is ~0.5772. The correction factor $\left(\frac{\gamma}{2 \ln N}\right)$ in (9.8) decreases in value as N the number of waves in a wave record increases. For a typical wave record of 20-minute duration, there will be about one hundred 12-second waves or about two hundred 6-second waves. For these values of N(i.e. N = 100 or 200), the correction factor is about 0.05 thus giving an error of about 5 percent. For practical purposes, this correction factor can be ignored and equation (9.7) can be used to estimate the probable maximum wave height. The following examples will illustrate the use of equations (9.4) through (9.8)

Example 1: Given a sea state for which H_s = 5 m; what is the probability of observing waves higher than 6 m?

Solution: From (9.4), we can write the probability that an observed wave height will exceed a given value H_1 as

$$p(H_1) = \exp[-2(H_1/H_s)^2]$$

Given, $H_1 = 6$ m and $H_s = 5$ m

$$\therefore\ p(H \geq 6\ m) = \exp[-2(6/5)^2] = 0.06 \qquad (9.9)$$

Example 2: Estimate the significant wave height for the wave record of Figure 4.1a

Solution: Use equation (9.6). Take $H_1 = 4$ m. An inspection of the wave record shows that there are about 13 waves whose amplitudes exceed 4 m and that there are a total of about 106 waves on the wave record. Taking n = 13, N = 106 and substituting on the right hand side of equation (9.6) gives 3.9 m; since this is the significant amplitude we double this value to obtain the significant wave height as 7.8 m. The spectrum plot for the wave record of Figure 4.1a is shown in Figure 9.1 for which the significant wave height using the FFT algorithm comes out to be 7.2 m. Our estimate of H_S using equation (9.6) is quite close to the value calculated using the FFT algorithm.

Example 3: What is the probable maximum wave height for the wave record of Figure 4.1a?

Solution: The wave record has a total of about 106 waves. The significant wave height for this record is 7.2 m as obtained by the FFT algorithm. Using equation (9.7), we have

$$H_{max} = H_s \sqrt{\frac{\ln N}{2}}$$

$$= 11.0\ m \qquad (9.10)$$

The probable maximum wave height for the wave record in Figure 4.1a comes out to be 11.0 m.

Besides these examples, there are many other operational procedures for utilizing waverider spectral data in real-time environment. An interesting example is provided by Britton(1981) which demonstrates how the three-hourly spectral wave data from an offshore buoy in northeast Pacific can be used in conjunction with synoptic weather charts to monitor the movement of a depression approaching the Gulf of Alaska. The Marine Weather Service of the NOAA in Washington U.S.A. has developed a system of reporting spectral wave data from approximately 45 buoys off the north American continent through their AFOS (Automation of Field Operations and Services) program. These spectral data from the buoys are reported on the specially designed AFOS

spectrum consisting of 15 frequency bands. This spectral information
is transmitted in near-real time via the Global Telecommunication
System and can provide useful guidance on nearshore wave conditions
and swell arrival times.

The above examples discuss the utility of data from a waverider which measures only the heave signal and provides a frequency
plot similar to that in Fig. 9.1. The recently developed pitch-and-
roll buoy (also called WAVEC buoy) can provide a number of directional
parameters based on buoy's heave and slope signal analysis. A recent
paper by Kuik, Van Vledder and Holthuijsen (1988) describes a method
for routine analysis of the WAVEC buoy signal to yield four directional model-free parameters per frequency, namely the mean direction, the
directional width, skewness and kurtosis of the directional energy
distribution. Two additional parameters, the spectral weighted mean
direction and the unidirectivity index, which is a good indicator
of the bimodality in the directional wave spectrum, have also been
defined (see WMO, 1987). These six parameters can completely specify
the various characteristics of the directional spectrum. The variation
of these parameters with respect to frequency can be displayed in
several two-dimensional graphs. Typically, however, a WAVEC record is
analyzed to generate only a selected number of plots which can provide
useful information for most offshore and coastal applications. For
example, Figure 9.2 shows a one-dimensional energy density plot, variation of mean direction and directional spread and a polar contour
plot of the two-dimensional spectrum as obtained from a WAVEC buoy
which was installed off the Labrador coast during the LEWEX (Labrador
Extreme Wave EXperiment), March 1987. The Figure also shows the
significant wave height, peak period and the direction of maximum
spectral value. Such composite plots can provide useful information on
directional spectra.

9.3 Wave Products and Wave Climatology

Typically, an operational spectral wave model can provide a
number of wave products like significant wave height, primary and
secondary wave period as well as direction and a two-dimensional (frequency vs. direction) energy spectrum. Additional parameters like,
whitecapping, which is based on the percentage of breaking waves can
also be generated by operational wave models. As an example, a sample
energy spectrum from the spectral wave model ODGP is presented in
Table 9.I. The Table shows 360 spectral density values (in m^2/Hz)
generated by the 15 frequencies and 24 directions of the ODGP model at
a selected grid point near site 31b in the Canadian Atlantic (see Fig.
7.8 for location of the site). Also shown in the Table are wind speed
and direction at the grid point at which the spectrum is generated.

185

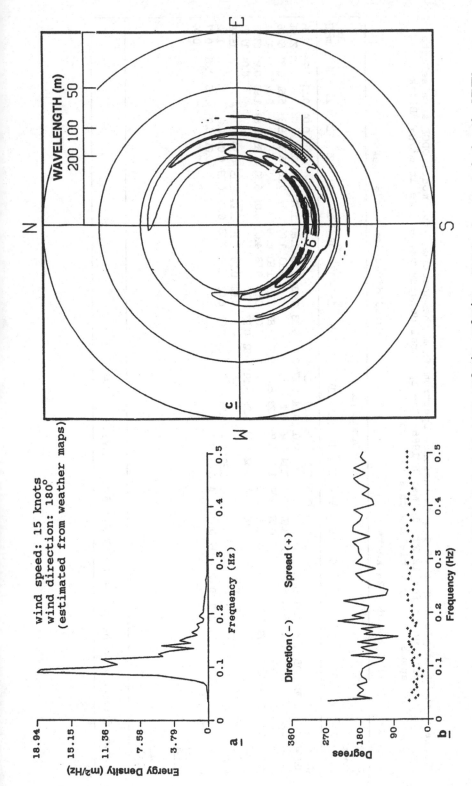

Figure 9.2: A sample output from a WAVEC buoy at station 258 (49°58'N, 47°37'W) off Labrador coast during the LEWEX measurements. Date: 15 March 1987; Time: 0315 GMT. a. one dimensional spectral plot, b. variation of mean direction and direction spread with frequency, c. polar contour plot of directional wave energy. For the above record, significant wave height: 3.3 m, peak period: 11.1 s, direction of spectral maximum: 175 deg. Direction convention is coming from (from archives of MEDS, Ottawa)

Table 9.I: A sample output of two-dimensional spectrum from the spectrum wave model ODGP at a selected grid point for 12GMT, 10 March 1986. The spectral density values are in m²/Hz. Blank spaces in the table indicate zero values.

wind speed : 28 knots   wind direction : 278 degrees

Freq. Hz	1	2	3	4	5	6	7	8	9	10	11	12	13	14	15	16	17	18	19	20	21	22	23	24	Sum m²/Hz
0.039								0.02	0.13		0.16	0.07	0.02									0.04	0.52	1.24	2.36
0.044	1.42	1.78	1.15	1.26	1.13	0.32					0.02	0.14	0.47					1.80	1.71	1.37	0.38				27.61
0.050	0.02	0.14	0.56	1.78	6.98	6.28	4.37		2.39	2.14	1.49	0.49		3.40	4.33	3.63		3.29	2.52	7.55	6.33	5.38	5.61	3.22	62.36
0.056	2.66	1.64	0.65					3.97	0.59	1.51	5.85						0.13	1.06				0.02		0.18	43.51
0.061	0.54	3.83	6.78	6.78	7.14	4.05	3.54	1.98	0.90			7.28	6.62	6.85	3.83	3.17	1.80	0.79	6.04	6.60	3.85	3.69	1.98	0.94	63.35
0.067								0.02	0.61	2.52				1.82	0.88	0.02	0.05	4.06	2.14	1.15	0.59			0.04	17.97
0.072	0.16	2.84	2.23	2.93	1.51	0.76					5.31	3.26	3.53												15.68
0.081					0.01				0.03	0.53	0.41	0.23			0.01	0.02	0.20	1.15	0.05	0.09	0.05	0.01			1.52
0.092			0.17				0.02	0.05	0.14	0.09	0.06					0.02	0.04	0.05			0.01	0.03	0.05	0.05	0.77
0.103	0.03	0.01			0.01		0.05	0.11																	0.04
0.117																									
0.133																									
0.158																									
0.208																									
0.308																									

Significant wave height = $4 \sqrt{\Sigma E(f_i)\Delta f_i}$ = $4\sqrt{1.3192}$ = 4.6 m

the significant wave height for this spectrum can be calculated by
summing up the spectral density values corresponding to each frequency
and using the formula

$$H_s = 4\sqrt{\Sigma E(f_i)\Delta f_i}$$

Here E(fi) is the total energy for the ith frequency band, Δf_i is the
width of that frequency band and $\Sigma E(f_i)\Delta f_i$ represents the total area
under the spectrum (this corresponds to the parameter m defined in
eq. 9.1). For the ODGP model, the width Δf_i varies from 1/180 Hz for
the first seven bands to 2/15 Hz for the last band. Summing over the
various frequency bands, we obtain the significant wave height value
as 4.6 m (this may be compared with the measured significant wave
height value of about 6.0 m at site 31b as displayed in Fig. 7.18).
Besides the significant wave height, the two-dimensional spectrum can
be displayed in terms of a frequency plot (frequency vs. energy) or
direction plots (direction vs. energy) for selected frequencies; such
plots can provide useful displays for operational purposes. Further,
the spectral density values of Table 9.I can be used to determine the
wave-induced ship motions through the Response Amplitude Operator
(RAO) which is defined as the ratio of the amplitudes of the individual motion responses to the amplitudes of the individual wave components. The use of RAO and related spectral wave information in seakeeping operations has been well documented by Bales(1987).

Besides ship operations, several other coastal and offshore
activities require a sound knowledge of wave products and wave
climatology either at a specific location or over a coastal or offshore region. Wave data collected at a waverider location can provide
a database useful for developing appropriate wave climatology and
wave statistics. As an example, Figure 9.3a shows a scatter diagram
of significant wave height versus peak period at Logy Bay (47.6°N,
52.5°W) off Newfoundland, Canada. The scatter diagram is based on a
total of 3235 wave records collected from 17 June 1976 to 9 January
1978. For each record, the significant wave height is determined from
eq. (4.11) where m is given by the area under the spectrum (eq. 9.1).
The other half of the Figure (9.3b) shows the peak period histogram
in which the ordinate is expressed in terms of percentages of occurrences. The scatter diagram together with the peak period histogram
can provide a useful picture of wave distribution at a location. For
example, at Logy Bay, the most frequently observed waves have a period
between 9 and 10 seconds, while only about three percent of total
observations have a wave period of 15 s or more. Further, the numbers
in the scatter diagrams (Figure 9.3a) can be converted to obtain
observed percentage of wave records for which the significant wave
height exceeded a certain value. These observed percentages can be

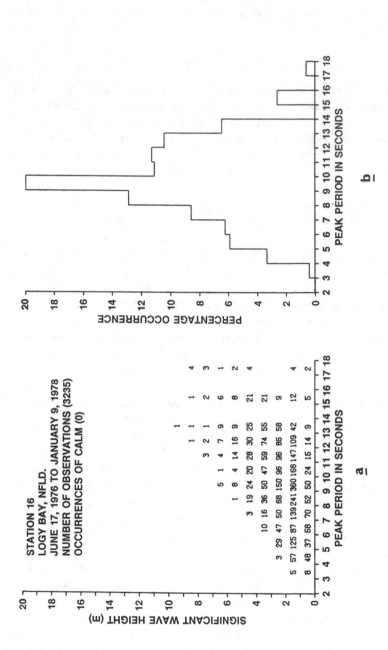

Figure 9.3: <u>a</u>. Scatter diagram for Logy Bay, Newfoundland, showing distribution of significant wave height versus peak period <u>b</u>. corresponding peak period histogram. also shown are the period of time over which observations were made and the total number of observations included in the preparation of the diagram. The 'occurrence of calm' represents the number of records for which the characteristic wave height is less 15 cm. For wave heights less than 15 cm, signal-to-noise ratio problems essentially render the period indeterminate (from the archives of MEDS, Ottawa).

plotted against wave heights to obtain an exceedence diagram as shown in Figure 9.4.

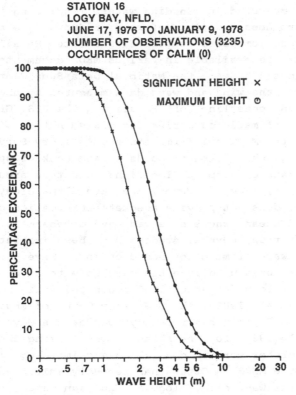

STATION 16
LOGY BAY, NFLD.
JUNE 17, 1976 TO JANUARY 9, 1978
NUMBER OF OBSERVATIONS (3235)
OCCURRENCES OF CALM (0)

SIGNIFICANT HEIGHT ×
MAXIMUM HEIGHT ⊙

Figure 9.4: Exceedence diagram showing the observed percentage of wave records for which the significant wave height exceeded a certain value. The curve labelled 'Maximum' is the exceedence curve for the most probable maximum wave height in a twenty-minute period. (from archives of MEDS, Ottawa)

Here the lower curve is drawn for the significant wave height while the upper curve (labelled maximum) can be interpreted as the exceedence curve for the most probable maximum wave in a twenty-minute period. This curve is computed using the Rayleigh distribution for the heights of sea waves in a narrow band of frequencies. Using this distribution, an expression for the probable maximum height (H_{max}) is obtained in terms of significant wave height (H_s) and the number of waves in the wave record (see eq. 9.7). The number of waves in a wave record is obtained by dividing the length of the record time (which is usually 20 minutes) by the peak period of the spectrum. The exceedence diagrams can provide useful information for planning of offshore development. From Fig. 9.3 it can be inferred that there is a 15 percent chance that the probable maximum wave height will exceed 5 m at Logy Bay.

Besides these products, there are a number of other wave products that describe the temporal and spatial distribution of important wave parameters. In Canada, the MEDS in Ottawa archives all marine environmental data measured in Canadian waters (including the Canadian Great Lakes) and produces a variety of wave products like surface elevation trace, spectrum plot, scatter diagram etc. Details of these wave products are available in various MEDS documents (see for ex. MEDS, 1981). In the U.S.A., the National Data Buoy Centre in Mississippi archives the wind and wave data measured by about 45 buoys (located off the U.S. coastline, off Hawaii and the U.S. Great Lakes) and prepares annual climatic summaries for dissemination of these wave data. In the U.K., the Marine Information and Advisory Service of the Institute of Oceanographic Sciences holds a data bank of spectral and non-spectral wave data of about 60 locations in waters around the British isles (Draper, 1984). However, the wave climatology based on observed data alone does not provide adequate information over wide open areas of world oceans where observed wave database is extremely limited. In view of this, several studies have been initiated to develop deep-water wave climatology based on model wave products. The U.S. Navy's SOWM has been used in a hindcast mode to develop a 20-year wind and wave climatology for about 1600 ocean points in northern hemisphere (Bales et al. 1982). The U.S. Army Corps of Engineers, Waterways Experiment Station has recently completed a wave hindcast study (Corson et al. 1981) for the Atlantic Ocean using the Resio (1981) model and further work on coastal wave climatology is in progress. In Canada, the MEDS in Ottawa has recently initiated a project to develop a spectral wave climatology in Canadian waters based on the ODGP model. Similar efforts have been initiated in many European countries to develop wave climatology in the North Sea and in other European waters.

9.4 Extreme Wave Statistics

The forecasting of extreme sea-state has been a subject of study for many statisticians and oceanographers. A knowledge of extreme sea-states can be very useful for ship routing as well as for planning of coastal and offshore development. A statistical analysis of wave heights using extreme-value distributions has been reported by Thom (1971), Borgman and Resio (1977), Forristall (1978) and LeBlond (1981) among others. A recent review paper (Muir and El-Shaarawi, 1986) summarizes the data limitations, statistical assumptions and distributions used in calculating extreme wave heights and related parameters. Forristall(1978) analyzed 116 hours of hurricane-generated wave data and found that a Weibull distribution expressed in the following form

$$p(H \geq H_1) = \exp[-\frac{1}{\beta}(H_1/\sqrt{m})^{\alpha}] \; ; \quad \alpha = 2.126 \; , \quad \beta = 8.42 \qquad (9.11)$$

gives a good fit to a large number of waves measured at 8 sites in the Gulf of Mexico. It may be noted that for $\alpha = 2$ and $\beta = 8$, the Weibull distribution reduces to the Rayleigh distribution (see eq. 5.9) which was used by Longuet-Higgins (1952) to obtain suitable expressions for the significant wave height and other wave parameters. Forristall's analysis indicates that the Rayleigh distribution substantially over-predicts the height of the highest waves in a wave record. Thom(1971) analyzed the extreme wave height data from 12 ocean station vessels using a Fisher-Tippet Type I distribution given by

$$P(H \leq x) = \exp[-\exp - \frac{(x - A)}{B}] \qquad (9.12)$$

Here A and B are the parameters of the distribution. This is the simplest of the three Fisher-Tippet type distributions. LeBlond (1981) also suggests the use of the simplest Type I distribution to analyze the hindcast data of extreme wave heights. Equations (9.11) and (9.12) give $p(H_1)$, the probability that wave height at a location will exceed a given value H_1. The probability $p(H_1)$ can be used to calculate the return period R (in years) for the extreme value by the formula

$$R = \frac{T}{1 - p(H_1)} \qquad (9.13)$$

Here T is the average value of the interval (in years) over which the distribution of the extreme values are sampled. For example, if the extreme values are obtained from a succession of 3-hourly wave re-cords, $T = 3$ hrs $= 3.42 \times 10^{-4}$ years; if a sample of extreme values consist of the largest (or highest) values from a succession of annual records, $T = 1$ year. The return period R in (9.13) can be interpreted as the average waiting time between occurrences of a wave with a height equal to or exceeding H_1. Equation (9.13) is quite useful in developing statistics of extreme wave conditions which a coastal or offshore structure is designed to survive. These conditions are called design wave conditions. The wave height associated with a 100-year return period is frequently used to determine the design of offshore structures. If we use $p(H_1) = 0.99$ and $T = 1$ year in (9.13), we obtain $R = 100$ years. Solving the probability distribution with $p(H_1) = 0.99$, we can obtain H_1 the expected maximum height in a 100-year period.

The above statistical methods, although useful, appear to give over-estimates of design wave parameters when compared with observed data (Earle, 1979). For reliable estimates, observational data base should therefore be considered as the most appropriate source of information. However, observed wave data are not available over long periods of time at many locations; accordingly, it becomes necessary to generate wave data by using wave models in a hindcast mode. Several recent studies have been initiated to develop a comprehensive wave database by running well-tuned operational wave models in a hindcast mode for a series of historical storm events. Two of these studies worth mentioning are the NESS (North European Storm Study) as reported by Francis(1987) and the recently initiated Environment Canada study on extreme wind and wave hindcast off the east coast of Canada (Swail, Cardone and Eid, 1989).

9.5 Real-time Wave Analysis and Prediction

Routine wave analysis and prediction either on a regional or on a hemispheric or global scale is being performed by meteorological and oceanographic services of many countries at present. The wave models used for real-time wave analysis and prediction vary from the classical nomogram-based procedures to the state-of-the-art spectral wave models. In the following, brief details of operational wave prediction procedures used in Canada, U.S.A., Europe and elsewhere are presented:

a. Canada: Wave charts in Canadian waters have been prepared and issued to the public since early 1970's by the Meteorology and Oceanography (METOC) centres of the Department of National Defense. These wave charts are based on a SMB-type nomogram and uses the continuity principle to advect the wave fields (Morgan,1971). These hand-analyzed wave charts are issued four times a day by the two METOC centres namely, Halifax (Nova Scotia) on the east coast and Esquimalt (British Columbia) on the west coast. A sample chart issued by the METOC centre in Halifax for northwest Atlantic is shown in Figure 9.5. The chart presents the sea-state over the northwest Atlantic during the passage of one of the CASP storms (IOP 8; see Fig. 7.16) which moved over the Scotian shelf on 15-16 February 1986. The elongated wave height maxima south of the Scotian shelf appears to be in response to the movement of the storm centre along the IOP 8 track as shown in Fig. 7.16. In general, the quality of these METOC charts has been found to be quite comparable to that produced by an operational spectral wave model; this has prompted the development of a computerized version of the SMB technique and has led to the 'Parametric Wave-Model' which is presently the operational wave model of the Atmospheric Environment Service (AES), Environment Canada. The governing equations for the parametric wave model are expressed as

$$
\begin{aligned}
\text{wave height} \quad H &= 0.283g^{-1} U^2 \tanh(0.0125 X^{0.42}) \\
\text{wave period} \quad T &= 7.54g^{-1} U \tanh (0.077 X^{0.25})
\end{aligned}
\tag{9.14}
$$

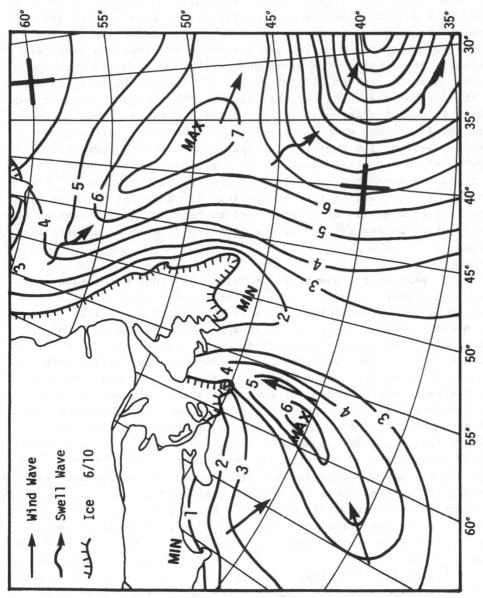

Figure 9.5: A significant wave height chart prepared by the METOC centre in Halifax, Nova Scotia for 16 February 1986, 00 GMT. Wave heights are in meters. The sea-ice boundary showing ice concentration greater than or equal to 0.6 is also shown

In these equations, $X = gU^{-2}x$ is a nondimensional fetch composed of a constant wind speed U, the total fetch x and g the gravitational acceleration (It may be noted that eq. 9.14 is the same as eq. 4.7 with parameters A_1, A_2 etc. appropriately defined). The pair of eq. 9.14 is applied to a suitable oceanic region consisting of several target points at which forecasts of significant height, major swell height, combined wave height and period and direction of waves and swell are calculated. In computing the waves and swells, 24 rays are extended from each target point at 15° intervals. The parametric model operates over the northwest Atlantic as well as over the northeast Pacific and Figure 9.6 shows the Atlantic and the Pacific domains of the parametric model together with the target points. At each of the target points, a 24-direction ray grid is superimposed which represents a discrete set of potential lines of fetch for each target point. The rays terminate when they reach land or the ray grid boundaries shown in Figure 9.7 for each of the two domains. For the Atlantic domain, the sea-ice boundary is also recognized by terminating the rays at the ice-edge which is represented by the boundary of 6/10 or greater ice coverage.

The parametric wave model is driven by winds provided by the CMC (Canadian Meteorological Centre) weather prediction model at 1000 mb; these winds are generated on a 42.3 km grid at six-hourly intervals. A cubic spline interpolation is applied to obtain winds at any given point in space and time along each ray. Since the wave model is interested only in the wind components along the ray, wind speeds are set to zero if the wind angle differs by more than 25 degreees from the ray direction. The potential wave generation regions are then defined as regions (in space and time) along each ray where the wind speed component is greater than 5.14 ms^{-1} (10 kt). A moving fetch procedure (Wilson, 1955) is used to take into account both the space and time variability of the wind. The equations (9.14) are integrated along each ray using a stepwise numerical integration procedure which assumes a steady wind speed (along each ray) over a small fetch (up to 74 km) and for a short duration (up to 2 h). A simple assumption based on wind speed values allows the calculation of wind waves and swell waves arriving at a target point (see Macdonald and Clodman, 1987). A sample wave height chart generated by the parametric wave model is shown in Figure 9.8; this chart corresponds to 16 February 1986, 00Z and can be compared with the hand analyzed METOC wave height chart (Fig. 9.5) which corresponds to the same date and time. In general, there is a good agreement between the two wave charts especially with respect to the shape of the major wave field patterns; thus the eq. 9.14 can provide a reasonable estimate of the wave height fields in a diagnostic sense provided the wind field is suitably specified. At present, the parametric model is run twice daily and wave forecasts

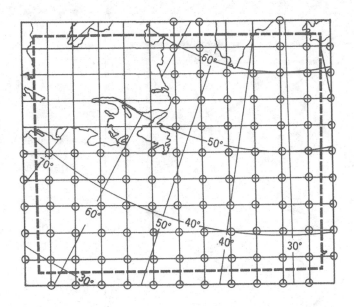

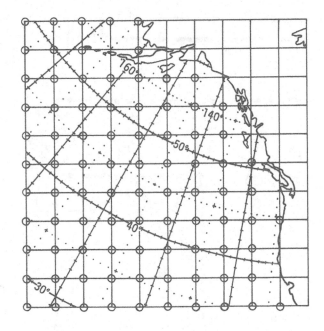

Figure 9.6: The Atlantic and the Pacific domain of the Parametric Wave Model, the present operational wave model of the Atmospheric Environment Service (AES), Canada. The target points are shown by circles where sea-state conditions are modelled using equations (9.14)

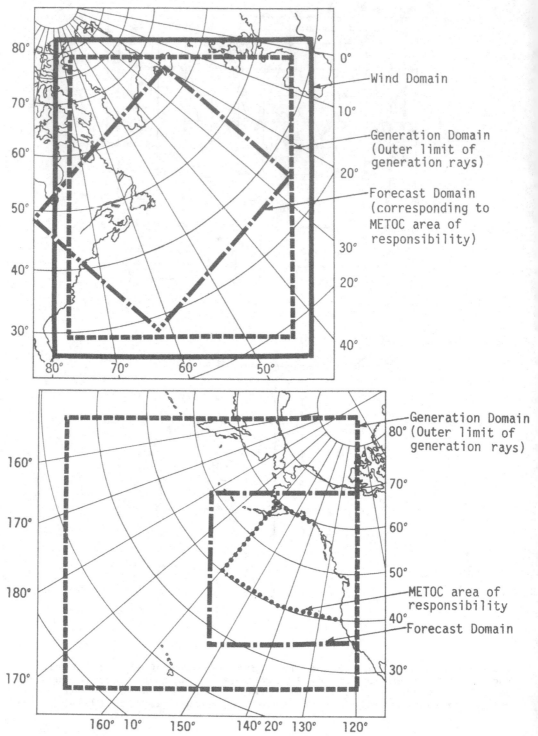

Figure 9.7: The wind domain and the wave generation domain for the Atlantic and the Pacific grid of the Parametric Wave Model

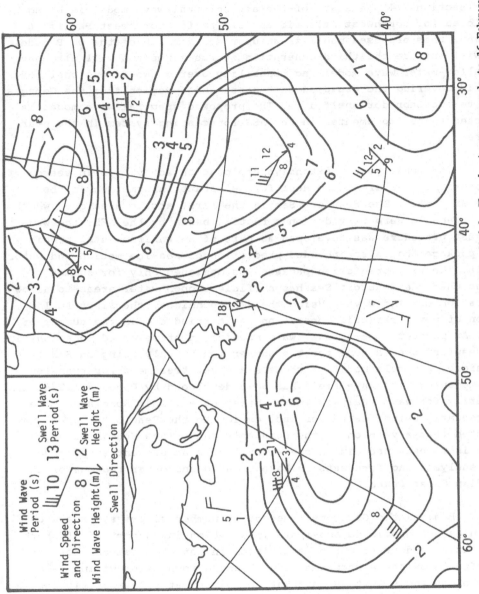

Figure 9.8: A wave height chart generated by the AES Parametric Wave Model. The chart corresponds to 16 February 1986, 00 GMT and can be compared with the corresponding METOC wave chart of Fig. 9.5. The wave and swell information using the wave plot model (see inset) is shown at selected locations only.

up to 36 hours are issued over the Atlantic and the Pacific domains.
Based on the results of the wave model intercomparison study during
CASP (section 7.4), a state-of-the-art spectral wave model is being
developed for northwest Atlantic as well as for northeast Pacific at
present; this proposed model is an upgraded version of the ODGP model
and will include the third-generation source terms and will also have
a shallow-water wave modelling capability over a two-dimensional nested grid covering the Canadian east coast shelf region from the Georges
Bank to the Labrador shelf area. The proposed spectral wave model is
expected to be implemented in the AES forecasting system in the near
future.

In recent years, extensive exploration activities in search of
hydrocarbons have created a need for sea-state information in the
Canadian Arctic. The Beaufort Sea is the first arctic area over which
wave forecasts were provided on a routine basis by the AES; the wave
forecast procedure was developed as part of an environmental prediction package that also included prediction of sea-ice motion and water
levels. The wave prediction package was designed only for summer
months when the Beaufort Sea has sufficient open water areas to allow
generation of wind waves. Using the SMB technique, the first full
season of wave prediction operation was carried out in the summer of
1976. At present, operational wave prediction at selected locations in
the Beaufort Sea is done for the summer months only using an SMB type
technique. A simple model which uses a novel treatment for transfer of
wind momentum into wave field has been developed by Donelan (1978)
primarily for wave analysis in the Great Lakes. This Donelan's model
is presently being tested for application to the Beaufort Sea. It may
be noted that a version of the Donelan's model (see Schwab et al.
1984) is already used in an operational mode at present to provide
wave analysis and forecasts in Lake Ontario and other regions of the
Canadian Great Lakes.

Besides the Government sector, a number of private research
companies are providing sea-state information in support of offshore
exploration and other marine activities in Canadian waters. Most of
the efforts by these companies are directed towards providing wave
climatology and sea-state information, either at site specific locations or over a localized area.

b. U.S.A: Wave analysis and wave prediction activity in the
U.S. Government sectors are mainly with three Federal Government agen-
cies, namely, the U.S. Army, the U.S. Navy and the NOAA (National
Oceanic and Atmospheric Administration) in Washington, D.C. The U.S.
Navy has been operating the SOWM (Spectral Ocean Wave Model) for the
three ocean basins of the northern hemisphere since about 1974. The

SOWM has been replaced since June 1985 by its global version called
the GSOWM which may be considered as the first global operational
spectral ocean wave model. The GSOWM is operated at the U.S. Navy's
Fleet Numerical Oceanography Centre in Monterey, California and is
driven by the NOGAPS (Navy Operational Global Atmospheric System)
which provides surface winds applicable at 19.5 m level and adjusted
for vertical thermal stratification. The evaluation of GSOWM against
SOWM has been discussed earlier in section 7.2; the error statistics
presented in Table 7.I shows that the GSOWM reduces the RMS errors in
wave heights by up to 30 percent when compared against SOWM. The U.S.
Army has primary responsibility to develop and provide wave analysis
and wave information service in the nearshore and coastal regions of
the U.S.A.. Most of the U.S. Army's wave analysis work relates to the
development of spectral wave climatology for the continental shelf of
the United States with inclusion of bathymetry and shallow-water effects like wave refraction and bottom friction. The U.S. Army operates
a verison of the Resio model applicable to waters of arbitrary depth;
the model is primarily used in a hindcast mode for assessment of wave
observations and related work. The NOAA in Washington has recently
(since early 1986) put into operation a spectral wave model covering
global oceans between 75°N and 70°S; the NOAA wave Model is an upgraded version of the SAIL model described by Greenwood et al (1985). The
NOAA wave model is presently driven by winds extracted from the lowest
σ-level of their operational weather prediction model and brought down
to 10 m level using a simple logarithmic profile. Using the NDBC
(National Data Buoy Centre) buoy data, the NOAA wave model products
have been evaluated in conjunction with the GSOWM wave products and
the NOAA model appears to be underforecasting the significant wave
heights slightly where the GSOWM tends to overforecast (Esteva and
Chin, 1987). A recent study by Esteva (1988) has demonstrated the
utility of incorporating satellite-sensed wave height data to obtain
improvement in the NOAA wave model output. The use of wave height data
from the presently orbitting satellite GEOSAT for real-time wave
analysis and forecasting is being experimented at present. Besides
the global wave model, the NOAA also operates a regional scale model
covering the Gulf of Mexico in which shallow water effects, namely
wave refraction and bottom friction are included; preliminary results
on the evaluation of this Gulf of Mexico model have been recently
reported by Chao(1989).

Besides the Government sector, there are a number of private
companies which are actively engaged in providing wave analysis and
sea-state information services for a variety of purposes like offshore
exploration, international shipping etc.

c. Europe: Several countries in western Europe presently operate a second generation spectral wave model belonging either to the
CH (Coupled Hybrid) or the the CD (Coupled Discrete) class. Most of
these models operate over European waters including the North Sea, the
northeast Atlantic and the Mediterranean Sea. Some of the operational
models of Europe have already been referred to in Chapter 5. For example, GONO is the present operational model in the Netherlands, while
HYPAS is the operational model in West Germany; both GONO and HYPAS
have grids which cover the North Sea and parts of the northeast
Atlantic. The British meteorological office presently operates a
global spectral wave model which is an upgraded version of the Golding
(1983) model; the BMO model operates on a global (Fig. 5.11) as well
as on a European (Fig. 5.12) grid as discussed earlier. The ECMWF in
Reading, U.K., is currently testing the global WAM model on a 3° x 3°
grid using the 10 m level winds extracted from the ECMWF weather prediction model; the global WAM model is expected to be put in an operational mode at the ECMWF by the end of 1989. The Norwegian Meteorological Institute operates a Coupled Hybrid model called NOWAMO which
was developed originally in 1966/67 by Odd Haug (see Guddal, 1985);
the NOWAMO is a deep-water model and presently operates over a grid
covering northeast Atlantic with a grid spacing of 150 km. A discrete
spectral wave model based on the PTB (Pierson, Tick, Baer) formulation
has been developed by the Danish Hydraulics Institute, Denmark and a
version of this model called DHI-System 20 is being used for real-time
wave analysis and forecasting in the Danish North Sea and vicinity; a
shallow-water verison of the DHI-System 20 has also been developed
and has been used to develop wave climatology on the west coast of
Denmark. In France, an upgraded version of the classical DSA spectral
model called DSA-5 has been in operation since about 1970. The model
operates on a 30 x 50 grid mesh covering a major portion of the north
Atlantic and has a grid spacing of 90 n mi everyhwere; the model is
presently operated by the National Meteorological Directorate in conjunction with the French weather forecasting service. Recent studies
in France on the testing of the north Atlantic verison of WAM model
have been reported by Guillaume(1988). A shallow-water wave model for
the Adriatic Sea with specific applications to the Italian coast has
been developed by Cavaleri and Rizzoli(1981). The model is used in
Italy for applications to coastal areas in the north Adriatic Sea.

d. Elsewhere: A coupled discrete wave model MRI-II (Uji, 1984)
is presently the operational wave model of the Japan Meteorological
Agency. The model is an upgraded version of the MRI (Meteorological
Research Institute) model which was one of the ten SWAMP models. The
model operates over the northwest Pacific and uses a grid spacing
of 381 km. The model is driven by winds obtainable from the four-
layer northern hemispheric weather prediction model of the Japan

Meteorological Agency. The Australian Bureau of Meteorological Service
initiated a sea-state information service since about 1973 and at
present the Bureau operates a first generation hemispheric spectral
wave model covering the southern hemisphere south of 15°S. There is
also an Australian Regional Wave Model which covers an area from about
10°S to 50°S and from about 90°E to 180°E. Both these models are
driven by 10 m winds which are obtained using a marine boundary-layer
formulation to the surface winds generated by the hemispheric and
regional weather prediction systems of the Australian Bureau of
Meteorology. In New Zealand, a spectral wave model belonging to the DP
catagory with only deep water physics is used in operational mode at
present. The model grid covers the southwest Pacific including the
Tasman Sea and the Southern Ocean as far west as 120°E. The model uses
12 directions and 10 frequencies and the source terms, including the
nonlinear wave-wave interactions, are specified in a single function
(see Laing, 1983). In Malaysia, a modified GONO model is used
operationally over the South China Sea and the straits of Malacca.
The Phillipine Atmospheric, Geophysical and Astronomical Services
Administration (PAGASA) uses analyzed surface charts and meteorological satellite imagery to predict sea-states based on the significant
weather systems affecting various shipping zones. In addition, the
classical PNJ technique is used to provide a numerical guidance. Elsewhere, wave prediction techniques based on Bretschneider nomograms are
still in operational use at present.

In summary, wave models ranging from the classical nomogram-
based procedures to the most sophisticated spectral wave models are
in operational use at present. With increasing marine and offshore
activities, there is expected to be an increasing need of wave models
providing detailed sea-state information over coastal as well as offshore regions of world oceans.

REFERENCES

Allen, W.T.R. 1983: Loc. cit. (Ch. 4)

Bales, S.L., 1987: Practical seakeeping using directional wave
spectra. Proc. Symp. Measuring Ocean Waves from space Johns Hopkins
Univ. 15-17 April 1986, Johns Hopkins's APL Technical Digest, Vol. 8,
No. 1, 42-47.

Bales, S.L., W.E. Commins and E.N. Comstock, 1982: Potential impact of
twenty year hindcast wind and wave climatology. Marine Technology, 19,
119-139.

Borgman, L.E. and D.T. Resio, 1977: Extremal prediction in wave
climatology. Proc. of the Ports' 77 Conf., Vol. 1, ASCE, New York,
394-412.

Britton, G.P., 1981: An introduction to sea-state forecasting. U.S. Dept. of Commerce, NOAA, National Weather Service, Washington, 208 pp.

Cavaleri, L. and P.M. Rizzoli, 1981: Loc. cit. (Ch. 6)

Chao, Y.Y., 1989: An operational spectral wave forecasting model for the Gulf of Mexico. Preprints, 2nd International Workshop on wave Hindcasting and Forecasting, Vancouver, April 25-28, 1989. Environment Canada, 240-247.

Cooley, J.W. and J.W. Tukey, 1965: An algorithm for the machine calculation of complex Fourier series. Mathematics of Computation, XIX, No. 89-92, 297-301.

Corson, W.D., et al. 1981: Atlantic coast hindcast; deep water significant wave information. U.S. Army Corps of Engineers, Waterways Experiment Station. Wave Info. Study for U.S. Coastlines, WIS Rep. 2, 856 pp.

Donelan, M., 1978: A simple numerical model for wave and wind stress prediction. Unpub. Manuscript, Env. Canada, National Water Research Institute, Burlington, Ont., 11 pp + Tables and Figures.

Draper, L. 1984: The need for, and the provision of, instrumentally obtained wave climate data. Int'l Symp. Wave and Wind Climate Worldwide, London, U.K. April 12-13, 1984. The Royal Inst. of naval Architects, Paper No. 2.

Esteva, D.C., 1988: Evaluation of preliminary experiments assimilating SEASAT significant wave heights into a spectral wave model J. Geophysical Research, 93, C11, 14099-14105.

Esteva, D. and H. Chin, 1987: Loc. cit. (Ch. 5)

Forristall, G.Z., 1978: On the statistical distribution of wave heights in a storm. J. Geophysical Research, 83, 2353-2358.

Francis, P.E., 1987: The North European Storm Study (NESS). Proc. Int'l Workshop on Wave Hindcasting and Forecasting, Halifax, Nova Scotia, Sept. 23-26, 1986. Environmental Studies Revolving Fund, Report Series No. 065, Ottawa, 370 pp., 295-302.

Golding, B. 1983: Loc. cit. (Ch. 5)

Greenwood, J.A., V.J. Cardone and L.M. Lawson, 1985: Loc. cit. (Ch. 5)

Guddal, J., 1985: The Norwagian wave model NOWAMO. Ch. 17, Ocean Wave Modelling. The SWAMP Group, Plenum, 187-191.

Guillaume, A. 1988: The use of pitch-roll-heave data to investigate wave model performances - Application to the North Atlantic version of the WAM model. Presented at the Sixth WAM meeting, Paris, France, March 21-24, 1988.

Kuik, A.J. G.PH. Van Vledder and L.H. Holthuijsen, 1988: A method for routine analysis of pitch-and-roll buoy wave data. J. Physical Oceanography, 18, 1020-1034.

Laing, A.K., 1983: A numerical ocean wave model for the southwest Pacific. New Zealand Journal of Marine and Freshwater Research, 17, 83-98.

LeBlond, P. 1981: On the forecasting of extreme sea states. Marine Env. Data Service, Contract Rept. Ser. 2, Ottawa, Canada, 26 pp.

Longuet-Higgins, 1952: Loc. cit. (Ch. 4)

Macdonald, K.A., and S. Clodman, 1987: The AES parametric ocean-wave forecast system. Proc. Int'l Workshop on Wave Hindcasting and forecasting, Halifax, Nova Scotia, Sept. 23-26, 1986, Environmental Studies Revolving Fund, Rept. Series No. 065, Ottawa, 119-132.

MEDS, 1981: Marine Environmental Data Service, User Guide. Dept. of Fisheries and Oceans, Ottawa, Canada.

Morgan, M.R. 1971: The analysis and forecasting of sea and swell conditions in deep water. Tech. Memo., TEC 763, Atmospheric Env. Service, Downsview, Ont., 32 pp.

Muir, L.R. and A.H. El-Shaarawi, 1986: On the calculation of extreme wave heights: A review. Ocean Engineering, 13, 93-118.

Pierson, W.J. and W. Marks, 1952: The power spectrum analysis of ocean wave records. Trans. American Geophysical Union, 33, 834-844.

Resio, D.T., 1981: Loc. cit. (Ch. 5)

Schwab, D., J. Bennett, P. Liu and M. Donelan, 1984: Application of a simple wave prediction model to Lake Erie. J. Geophysical Research, 89, 3586-3592.

Swail, V.R., V.J. Cardone and B.M. Eid, 1989: An extreme wind and wave hindcast off the east coast of Canada. Preprints, Second Int'l workshop on wave hindcasting and forecasting, Vancouver, British Columbia, April 25-28, 1989, Environment Canada, 151-160.

Thom, H.C.S. 1971: Asymptotic extreme-value distributions of wave heights in the open ocean. J. Marine Research, 29, 19-27.

Uji, T., 1984: A coupled discrete wave model MRI-II. J. Oceanographical Soc. of Japan, 40, 303-313.

Wilson, B.W., 1955: Loc. cit. (Ch. 4)

WMO, 1987: A global survey on the need for and application of directional wave information. WMO/TD. No. 209, (S. Barstow and J. Guddal), World Meteorological Organization, Marine Meteorological and related Oceangraphic activities, Report No. 19, 1987, 94 pp.

CHAPTER 10
SUMMARY AND FUTURE OUTLOOK

10.1 General Comments

The objective of the monograph was to review the development of ocean wind-wave analysis and prediction and to provide an up-to-date status of wave analysis and wave prediction services that have been developed in many countries. A number of wave prediction models reported in recent literature has been summarized in this monograph; these models are presently used in research or operational modes in different parts of the world. The earlier Chapters of the monograph include some aspects of basic wave dynamics, wave generation, wave propagation and wave dissipation.

Since the pioneering development of Sverdrup and Munk, significant advances have been made in understanding and observing the physical processess affecting wind wave generation and growth. Theoretical studies of Miles and Phillips in the late fifties have provided suitable expressions for modelling the wind input term S_{in}. The international field experiment JONSWAP (in 1968, 1969) has provided a valuable insight into some of the physical processes of wave generation and wave energy redistribution; this has led to the development of parametric and hybrid models which incorporate the nonlinear energy transfer process ($s_{n\ell}$) in the wind-sea region. The field experiment by Snyder and his co-workers (Snyder et al. 1981) has helped provide a more accurate expression for the wind input term S_{in}. Besides this, several wave modelling efforts have helped establish the basic framework of contemporary wave models in the spectral energy balance equation as expressed by (5.1). Our knowledge of the dissipation terms S_{ds} still remains very limited and mainly qualitative. Pierson and his associates have developed, in their models, empirical expressions for wave dissipation based on the wave breaking mechanism. The study of Hasselmann relating the effect of whitecapping on the spectral energy balance and a recent extension of this study by Phillips has provided an improved understanding of wave breaking and dissipation processes. However, these and other dissipation processes like interaction of waves with surface currents and with bottom topography have not been accurately modelled so far, mostly due to insufficient observations. It is hoped that future observational programs may allow improved quantitative formulation of these processes for inclusion in the wave prediction models.

10.2 Present Status of Operational Wave Models

Most of the wave models used in operational mode at present
are either first or second generation spectral wave models. Most of
these models are currently operated at various national meteorological
centres and are generally driven by winds obtainable from a weather
prediction model. At most national centres the weather prediction
models are run so as to provide weather analysis and forecasts at
specified international times (ex. 00GMT, 06GMT etc.). The analysis
and forecast products from the weather prediction models are then
extracted for driving an ocean wave model. Private companies which
operate their own ocean wave models can prepare their own wind input
based on weather products received from national centres. Operational
wave models of many countries include shallow-water wave calculations
at present. Due to relatively larger computational requirements, the
shallow-water wave calculations are done either infrequently or on an
optional basis at many centres. Limited availability of computer power
in real-time environment is still a major factor in implementation of
upgraded versions of ocean wave models for operational use.

Appropriate validation of operational wave models is an on-
going task carried out at most national centres. A number of wave
model evaluation studies summarized earlier has provided a degree of
confidence in the use of operational wave products. A revealing test
of an ocean wave prediction model is its ability to specify the directional wave spectrum for realistic wind fields that vary in space and
time. So far, wave models have not been tested adequately for their
ability to specify the directional wave spectrum due to a lack of
sufficient wave directional data. Hopefully, technological advances in
the near future will lead to the development and inexpensive deploy-
ment of a system capable of measuring the directional wave spectrum
over many areas of world oceans; this will provide a much needed
directional wave database to assess the performance of a wave model. A
simple microwave radar technology for measuring directional wave
spectra has been recently described by Jackson, Walton and Baker
(1985); this technology has been validated at 10 km aircraft altitude
and appears to have a potential to provide global measurements of
directional wave spectra via a satellite.

10.3 Wave Model Initialization

Unlike numerical weather prediction models, an ocean wave
model is not initialized with observed wave data. Typically, a wave
model starts with zero wave energy everywhere (often identified as a
'cold start') and is driven with hindcast winds for a period of 24 to
48 hours. The 'spin-up' time of 1 to 2 days is required to bring up

the wave energy to realistic levels everywhere. Recent advances in
satellite technology have created a possibility of utilizing remotely
sensed wind and wave data for real-time wave analysis and forecasting.
Two aspects of wave model initialization are being investigated at
present. In one, the use of satellite-sensed winds over oceans are
being suitably blended with conventional wind data to produce an
initial wind input for driving an ocean wave model. Cardone (1983)
assimilated the wind data generated by a scatterometer aboard the
short-lived satellite SEASAT (28 June - 10 October, 1978) with the
conventional ship winds using a subjective analysis scheme. Cardone's
study was extended by Lalbeharry (1988) who developed an objective
analysis scheme to blend the SASS (SEASAT-A Satellite Scatterometer)
winds with the ship winds using appropriate weighting factors.
Lalbeharry's study further demonstrates how such a blended wind field
can provide an improved wind input to a spectral wave model and this
in turn can provide improved wave products in analysis as well as in
forecast mode.

The second aspect of the wave model initialization which is
being studied vigorously at present is the utilization of satellite-
derived wave height and wave spectrum data to initialize the model
wave field. A study by Thomas (1988) used wind and wave height data
from three waverider locations in the North Sea to assimilate into a
wave model, while Janssen et al (1989) used SEASAT wind and wave
height data to assimilate into the third generation WAM model. Similar
other studies, notably by Hasselmann et al (1988) are aimed at
developing a suitable algorithm to assimilate altimeter wave height
and SAR (Synthetic Aperture Radar) image spectra into a wave model;
these studies are geared towards utilization of wind and wave products
which would become available with the launching of the European Space
Agency (ESA) satellite ERS-1, by the end of 1990.

10.4 Wave Modelling in Shallow Water and in Nearshore Region

As mentioned earlier, wave modelling in shallow water is
receiving increasing attention at present. Operational wave models of
many countries include shallow-water wave modelling capabilities.
Typically shallow water effects are calculated on a nested fine-scale
grid and use, as boundary conditions, the deep-water model calculations. A number of nearshore and offshore operations require site-
specific sea-state conditions; these can often be provided by considering the sea-state information at the nearest deep-water location
provided by an operational wave model and then modifying the same for
local influences like bathymetry, bottom friction, percolation etc.
Several marine activities at present have become quite sophisticated
so as to require a detailed sea-state information over a localized

region like a bay area, a harbour entrance or a fishing and recrea-
tional zone. In view of these needs, wave models over localized nearshore areas have been developed and put into operational use. The
model developed by Holthuijsen and Booij (1987) applicable to river
estuaries and harbour entrances, is an excellent example of such a
wave modelling effort. With expanding marine activities over many
coastal regions of the world, efforts on nearshore wave modelling and
on some of the related topics like storm surges, sediment transport
etc. are expected to increase in the near future.

10.5 Future Outlook

The next major thrust on ocean wave modelling is expected to
be provided by the satellite-derived database on winds over ocean and
ocean wave spectra. Efforts to blend satellite-derived winds with
conventional data (ship winds, buoy winds) have already demonstrated
the utility of such a procedure to provide an improved wind specification over oceanic regions. With improved communication technology, the
procedure could be used in a real-time environment to provide improved
wind fields for driving ocean wave models. The measurement of ocean
wave spectra by instruments like SAR(Synthetic Aperture Radar), and
ROWS (Radar Ocean Wave Spectrometer) have been reported by Beal(1987),
and Jackson(1987) among others. The technology involved in these
instrumentations is quite sophisticated and the ocean surface products
that these instruments can provide appear quite impressive and promis-
ing. The impact of these products on ocean wave analysis and prediction is being investigated in a number of studies at present. By early
1990's, one or more satellites are expected to provide winds over
ocean and measurement of ocean wave spectra in real to near-real time;
this could provide an impetus towards significant improvement in ocean
wave analysis and prediction.

Finally, the importance of carefully designed field experiments on ocean wave measurements can not be underestimated. The
JONSWAP field experiment provided a valuable framework for wave model
research and development. Two more field experiments are planned
over the next three years. A synoptic experiment to study the two-
dimensional evolution of the surface gravity wave field is being
co-ordinated by Dr. R. Synder (U.S.A.) in the Bight of Abaco, Grand
Bahama Island during 1990 and 1991; another experiment called SWADE
(Surface WAve Dynamics Experiment) is being co-ordinated by Dr. M.
Donelan (Canada) to study the evolution of directional wave spectra
and will be conducted off the east coast of U.S.A. during the 1990/91
winter. These two field experiments are expected to provide a valuable
database for future wave modelling efforts.

REFERENCES

Beal R.C., 1987: Spectrasat: A hybrid ROWS/SAR approach to monitor ocean waves from space. Proc. Symp. Measuring Ocean Waves from Space, Johns Hopkins Univ. 15-17 April 1986, John Hopkins APL Technical Digest, 8, 107-115

Cardone, V.J., 1983: Potential impact of remote sensing data on sea-state analysis and prediction. Final Report, Oceanweather Inc., Submitted to Goddard Space Flight Centre, Greenbelt, Maryland (U.S.A.), December 1983, 95 pp.

Hasselmann, K. et al. 1988: Development of a satellite SAR image spectra and altimeter wave height data assimilation system for ERS-1. Max-Planck-Institut für Meteorologie, Report No. 19, Hamburg, July 1988 155 pp.

Holthuijsen, L.H. and N. Booij, 1987: A grid model for shallow water waves. Coastal Engineering, 1986 Proceedings, Ch. 20, 261-270, American Society of Civil Engineers, New York, U.S.A.

Jackson, F.C., 1987: The radar ocean wave spectrometer. Proc. Symp. Measuring Ocean Waves from Space, Johns Hopkins Univ. 15-17 April 1986, Johns Hopkins APL Technical Digest, 8, 116-127

Jackson, F.C., W.T. Walton and P.L. Baker, 1985: Aircraft and satellite measurement of ocean wave directional spectra using scanning beam microwave radars. J. Geophysical Research, 90, C1, 987-1004

Janssen, P., P. Lionello, M. Reistad and A. Hollingsworth, 1989: Hindcast and data assimilation studies with the WAM model during the SEASAT period. J. Geophysical Research, 94, 973-993.

Lalbeharry, R. 1988: Loc. cit. (Chapter 8).

Snyder, R.L. et al. 1981: Loc. cit. (Chapter 3).

Thomas, J.P., 1988: Retrieval of energy spectra from measured data for assimilation into a wave model. Q. J. Royal Meteor. Soc., 114, 781-800.

Coastal and Estuarine Studies
(formerly Lecture Notes on Coastal and Estuarine Studies)